Composite Materials

Edited by Mohammad Asaduzzaman Chowdhury, José Luis Rivera Armenta, Mohammed Muzibur Rahman, Abdullah Asiri and Inamuddin

Published in London, United Kingdom

IntechOpen

Supporting open minds since 2005

Composite Materials
http://dx.doi.org/10.5772/intechopen.87676
Edited by Mohammad Asaduzzaman Chowdhury, José Luis Rivera Armenta, Mohammed Muzibur Rahman, Abdullah Asiri and Inamuddin

Assistant to the Editors: Bengir Ahmed Shuvho

Contributors
Yanan Hou, Joseph Koo, Liguo Li, IfeOlorun Olofin, Ronggui Liu, Ihab Abdel-Latif, Charanjeet Singh, Sukhleen Bindra Narang, Doina Elena Gavrila, Victor Stoian, Alina Caramitu, Sorina Mitrea, Abdelkader Rahmouni, Mohammed Belbachir, Fatima Zohra Zeggai, Bachari Khaldoun, Redouane Chebout, Zahra Sayyar, Zahra Jamshidi, Veniamin Koshevoi, Anton Belorus, Anton Timchenko, Ilya Pleshanov, Roman Denisenko, Daniyar Sherimov, Ekaterina Vodkailo, Vijay Rangari, James Davis, Mohamed Abdelsabour Abdelsabour Fahmy

First published in London, United Kingdom, 2021 by IntechOpen
IntechOpen is the global imprint of INTECHOPEN LIMITED, registered in England and Wales, registration number: 11086078, 5 Princes Gate Court, London, SW7 2QJ, United Kingdom
Printed in Croatia

British Library Cataloguing-in-Publication Data
A catalogue record for this book is available from the British Library

Additional hard and PDF copies can be obtained from orders@intechopen.com

Composite Materials
Edited by Mohammad Asaduzzaman Chowdhury, José Luis Rivera Armenta, Mohammed Muzibur Rahman, Abdullah Asiri and Inamuddin
p. cm.
Print ISBN 978-1-78984-787-1
Online ISBN 978-1-83962-740-8
eBook (PDF) ISBN 978-1-83962-741-5

For EU product safety concerns:
IN TECH d.o.o., Prolaz Marije Krucifikse Kozulić 3, 51000 Rijeka, Croatia, info@intechopen.com or visit our website at intechopen.com.

Meet the editors

Mohammad Asaduzzaman Chowdhury is a Professor of Mechanical Engineering at Dhaka University of Engineering and Technology (DUET), Gazipur, Bangladesh. His research interests include engineering tribology, surface engineering, coating technology, polymer and composite materials, metals and alloys, erosive wear of different materials, material fabrication, characterization of materials, self–energy-generating materials, and atomic transfer mechanisms. He is an editorial board member for many international journals of repute. Currently, he is working as an editor and reviewer for many ISI and Scopus indexed international journals and books. He has also published many research and review papers in refereed international journals and conference proceedings. Dr. Chowdhury is a consultant, advisor, and expert member of many government and autonomous organizations. He has twenty years of teaching and research experience. His personal interests involve different cultural and social activities as well as writing articles, stories, lyrics, and poems for different newspapers and relevant media.

José Luis Rivera-Armenta has a BSc in Chemical Engineering, an MSc in Petroleum Technology and Petrochemicals, and a Ph.D. in Chemical Engineering, all from the Technological Institute of Madero City (ITCM), Mexico. Since 2003, he has been a full-time professor in postgraduate programs at ITCM and head of the thermal analysis, injection, and extrusion laboratory. He has been responsible for several research projects sponsored by Consejo Nacional de Ciencia y Tecnología (CONACYT) and the National Technological Institute of Mexico (TecNM). He has advised ten Ph.D., seventeen master's degrees, and five bachelor's degree theses. He has published fifty-five scientific articles, six book chapters, and has edited two books and one special issue journal. Dr. Rivera-Armenta is also an active reviewer for several journals.

Dr. Mohammed Muzibur Rahman received his BSc and MSc from Shahjalal University of Science & Technology, Sylhet, Bangladesh, in 1999 and 2001, respectively. He received his Ph.D. from the Chonbuk National University, South Korea, in 2007. After completing his Ph.D., Dr. Rahman worked as a postdoctoral fellow and assistant professor at pioneer research centers and universities located in South Korea, Japan, and Saudi Arabia (2007 to 2011). Presently, he is an associate professor at the Center of Excellence for Advanced Materials Research (CEAMR) and the Chemistry Department at King Abdulaziz University, Saudi Arabia, where he has been since 2011. He has published more than 300 international and domestic conferences and several book chapters and edited fourteen books. His research interests include composite materials, electrocatalysis, photocatalysis, semiconductors, nanotechnology, nanomaterials, nanoparticles, carbon nanotubes, sensors, ionic liquids, surface chemistry, electrochemistry, self-assembled monolayers, and devices.

Dr. Abdullah Mohamed Asiri received his Ph.D. from the University of Wales, College of Cardiff, UK, in 1995. He has been the head of the Chemistry Department at King Abdulaziz University since October 2009. He is also the founder and director of the Center of Excellence for Advanced Materials Research (CEAMR) as well as a professor of Organic Photochemistry. His research interests include chemistry, synthesis of novel photochromic and thermochromic systems, synthesis of novel coloring matters and dyeing of textiles, materials chemistry, nanochemistry, nanotechnology, polymers, and plastics. He is the editor-in-chief of the King Abdulaziz University Journal of Science. He is also a member of the editorial board of Pigments and Resin Technology (UK), Organic Chemistry in Sight (New Zealand), and Recent Patents on Materials Science (USA). Dr. Asiri is vice president of the Saudi Chemical Society (Western Province Branch).

Dr. Inamuddin is an assistant professor in the Chemistry Department, Faculty of Science, King Abdulaziz University, Jeddah, Saudi Arabia. He has extensive research experience in multidisciplinary fields of analytical chemistry, materials chemistry, and electrochemistry, and, more specifically, renewable energy and the environment. He has published 127 research articles in international journals of repute and eighteen book chapters and thirty-nine books published by renowned international publishers, such as Springer, Elsevier, Nova Science Publishers, Inc., CRC Press, and Taylor & Francis, among many others. He is a member of various journal editorial boards and serves as an associate editor for journals such as Environmental Chemistry Letter, Applied Water Science, Euro-Mediterranean Journal for Environmental Integration, and Springer Nature. He is also an editorial board member for Scientific Reports-Nature and editor for Eurasian Journal of Analytical Chemistry.

Contents

Preface

This book presents information about composite materials, which have a variety of applications in engineering and aeronautics, transportation, construction, sports, recreational activities, and so on.

Composite materials have anisotropic behavior, which makes these materials resistant to corrosion with a good weight-to-stiffness ratio. They are often reinforced with fibers or other particles and have the excellent capability for integration with other materials.

Composite materials consist of two or more constituents that have different physical and chemical properties. One of the constituents is the matrix, which can be polymer, ceramic, or metal. Among the most used matrix materials for composites are thermoplastic and thermoset polymers. Thermoplastics are the most common polymers and include polypropylene, polyvinyl chloride (PVC), polyethylene, nylon, polyethylene terephthalate (PET), acrylonitrile-styrene-butadiene (ABS), and others. These polymers are used in medical devices, tablets, vehicles, panels, frames, interior components, buildings, roads, air- and spacecraft, and so on.

Matrix materials can be metallic, such as aluminum, magnesium, and titanium, which can be used as catalysts as well as for energy storage and corrosion resistance. They can also be ceramic, including carbon, silicon carbide, and alumina. These materials can be used in tools, wear components, bioceramics, heat exchangers, coatings, and more.

This book reports on composite materials in two sections: “Composite Materials with Polymer Matrix” and “Composite Materials with Other Kind of Matrix.”

The first section evaluates the thermal and mechanical properties of thermoplastic and thermoset polymers reinforced with particles (SiC/SiO_2, metallic particles, etc.) and fibers. The section includes information on using the finite element method to analyze these materials at different temperatures, as well as discusses polymers such as polyaniline (PANI) and epoxy and their electrical and electronic properties and preparation methods.

The second section discusses new 2D composites such as thin films and porous silicon for biosensors devices, as well as modification of composites based on ferrite with Co^{2+} and Ti^{4+} and their conductivity and shielding properties.

In discussing the different materials, this book includes information on the design of the materials, their structure, and their preparation methods. Composite Materials is a compilation of works that reports on this expansive field and the extensive potential applications of composite material.

Dr. Mohammad Asaduzzaman Chowdhury
Department of Mechanical Engineering,
Dhaka University of Engineering and Technology (DUET),
Gazipur, Bangladesh

Dr. José Luis Rivera-Armenta
Tecnológico Nacional de México (TecNM)/
Instituto Tecnológico de Ciudad Madero,
México

Mohammed Muzibur Rahman, Abdullah Asiri and Inamuddin
King Abdulaziz University,
Jeddah, Kingdom of Saudi Arabia

Section 1

Composite Materials with Polymer Matrix

Chapter 1

The Role of Novel Composite of 2D Materials and Their Characterization, Properties, and Potential Applications in Different Fields

Zahra Sayyar and Zahra Jamshidi

Abstract

Thin layers or coating of transition metal dichalcogenides (TMDs) is a new class of two-dimensional (2D) inorganic materials with unique physical and chemical properties. This book chapter covers the recent research of thin film of 2D materials using various novel technologies to synthesize and grow monolayer 2D materials on different substrates in different fields based on the knowledge available in the literature. Thin film on substrate can be enhanced with the favorable properties. Therefore, selection of methods can play a key role in characterizing the coating. The novel coating processes on composite materials and their characterization, properties, and process and potential applications also have been discussed. The 2D materials that have been investigated created a thin film through different methods and were used to serve different biomedical purposes such as modifying drug release, improving energy efficiency and storing energy, catalysts, and so an.

Keywords: 2D materials, thin film, application, synthesize

1. Introduction

Nowadays, the potential of nanotechnology confirms us to use nanomaterials in all fields such as biotechnology, engineering, and medical; therefore, nanotechnology opens up new frontiers for innovation in these fields [1, 2]. There are diverse classes of nanomaterials in nanoscience, for example, the zero-dimensional (0D) class of quantum dots; one-dimensional (1D) class of nanoribbons, nanotubes, and nanowires; two-dimensional (2D) class of single-atom thick materials; and three-dimensional (3D) class of nanoballs and nanocones [1, 3]. Their dimensionality is important properties to distinguish these categories of nanostructure. If the same chemical elements or compounds have different dimensions, they will exhibit different properties.

Two-dimensional (2D) layered materials have attracted attention of researchers because of their unique electrical, mechanical, thermal, and optical properties [4, 5]. These types of materials include a few atoms or monomer units: thin sheets exhibiting covalent in-plane bonding and weak interlayer along with layer-substrate bonding resulting in unique chemical reactivity and several physical properties.

Graphene, hexagonal boron nitride (h-BN), and transition metal dichalcogenides (TMDs, e.g., MoS_2, $MoSe_2$, WS_2, and WSe_2) are kinds of these materials that they can exhibit high charge carrier mobilities, chemical inertness, and high mechanical strength [3, 5, 6]. In addition, 2D materials have the large specific surface area to use them in surface-active applications, for example, catalysis, sensing, separation, energy storage and conversion, and other related fields. There are various nanostructures such as metals, metal oxides, and metal chalcogenides that have established effective in reaching developed properties and creation of different purposes. Additionally, properties like atomic thickness and high anisotropy could be used to produce wearable electronic devices based on 2D materials due to their admirable mechanical flexibility and optical transparency, which these properties cause to develop 2D material-based electronic devices and wearable devices [7, 8].

In this chapter, we will mention the process of the synthesis and chemistry of synthetic elemental 2D materials and their application in different fields. Finally, we will offer perspectives and challenges for the future of this emerging field.

2. Methods of synthesis and chemistry of 2D thin film

Coating is the most widely used technology to enhance surface properties of substrates, and nanotechnology has played an important role in improving the coating action [9]. In the case of 2D materials, the monolayers can be arranged on top of one another in a selected arrangement to produce unique, heterogeneous multilayers, leading to van der Waals heterostructures [10]. Moreover, homogeneous multilayer materials, inorganic 2D materials, organic, and organometallic have also increased interest of researchers, which include 2D covalent organic frameworks and metal-organic frameworks. The structural varieties of organic ligands and the numerous linkage chemistries cause the best properties as desired composition, size, thickness, crystal phase, and surface property [7].

There are different synthetic methods based on "top-down" and "bottom-up" approaches for the synthesis of 2D materials. The "top-down" approaches contain the physical mechanical technique such as mechanical sonication, shear force, ion intercalation, and exchange. Despite a potential tool to produce large-scale single-layer or few-layer 2D materials at the ambient condition, this method has various disadvantages such as the broad distribution of sheet thickness and lateral size as well as low yield, low structure integrity, and high tendency of restacking. But, the "bottom-up" approaches, including chemical vapor deposition (CVD), physical vapor deposition (PVD), solid state reaction synthesis, and wet-chemistry synthesis, can overcome these disadvantages, and it can control the morphology and the structure of inorganic 2D materials. It also has numerous advantages such as up-scalability, solution processability, and eco-friendliness [6, 11–13].

2.1 Developed techniques of material production

Over the last decade, thin film of 2D materials such as grapheme has attracted worldwide research interest. However, there are many other materials in layered form with interlayer interactions, which made by van der Waals forces. Nowadays, these materials can be thinned with proper techniques to a few layers and even monolayers. Different methods can be used to create these layers, which are the following:

One of the main methods used to obtain monolayers is mechanical exfoliation. The process depends on the probability of cleaving the crystal of bulk materials and the bonding of metal. Therefore, only a monolayer is moved onto the desired substrate [14].

Method	Advantage	Disadvantage	Ref.
Sol-gel	Suitable to obtain high quality films up to micron thickness	To limit the use of metallic as subtracts for coating. To exhibit several drawbacks involving crack ability and thickness limits	[17]
Thermal spray (gas-flame, plasma)	To make nanocomposite coatings with a matrix of metal or alloy	High temperature	[18]
Cold spray	To allow fabrication of coatings at the lower temperatures with low porosity	High pressure	[19]
Spin coating	Preparation of polymer nanocomposite coatings. To create uniform thin films on substrates.	High speed	[20]
Electrodeposition	For the fabrication of nanocomposite coatings, which contain organic nanofillers	Formation of microcrack	[21]

Table 1.
Advantages and disadvantages of some methods.

There is a simple method for synthesizing 2D materials such as chemical vapor deposition (CVD). This method does not control the grown materials exactly and creates inhomogeneous films. Park group has recently established homogeneous, wafer-scale films like MoS_2 and WS_2 using metalorganic chemical vapor deposition (MOCVD). Therefore, they synthesized high-quality and large-area materials. Temperature, pressure, chemical reagents, and concentrations as experimental parameters can effect on the shape, size, morphology, and uniformity of the solution-grown nanostructures. Moreover, this method can control shape, size, uniformity, and layer thickness. Inappropriate conditions to produce colloidal solutions can cause agglomeration [15, 16]. Other technologies of the coating on composite materials recently concentrate great attention on researchers compared to **Table 1** with their distinct applications.

3. Characterization of 2D material coating

Coating texture, morphology, and appearance, particularly for viscosity and color, are important to characterize coating film. There are several nondestructive techniques to analyze and characterize these coatings in use, according to their film surface functionality. Extreme sensitivity and high magnification are involved due to the thin nature of films developed. In this section, to investigate morphology and texture of coating, some of the techniques are investigated. Electron microscopy techniques, such as scanning electron microscopy (SEM) and transmission electron microscopy (TEM), and the sensitive atomic force microscopy (AFM) can be used to evaluate morphology and texture of coating. AFM is a high-resolution imaging tool, which measures and manipulates matter at a nanoscale. It shows significant changes on the metal and coating surfaces. It offers 3D visualization, qualitative, and quantitative information on size, morphology, surface texture and roughness, statistical distribution, surface area, and volume distributions. TEM determines particle size and shape. Also, histograms of TEM images can be employed to count individual particles formed. Furthermore, TEM imaging has been employed to observe certain fabrication features between smart nanocontainers and inhibitors. SEM shows particle size and shape. Side view of the SEM images also provided dimensions of coating thickness on the brass substrate studied. UV-visible spectrophotometry is required for determining the reduction of metal ions to nanoparticles via surface plasmon resonance, which is also responsible for their unique colors upon reduction. Optical properties of the metallic nanoparticles are permissible. Fourier-transform infrared spectroscopy (FTIR) identifies possible molecules responsible for the reduction, stabilization, and capping of micro and nanoparticles coupled with coatings formed via IR absorption spectra. X-ray diffraction (XRD) determines particle crystallinity and nature of metallic particles in the coating types. Also, phase structure of metal nanoparticles was identified for some nanoparticles. Estimation of coating performance is usually employed in terms of coating degradation, corrosion kinetics, and electrochemical techniques. Scanning vibrating electrode technique (SVET) provides information on corrosion reactions initiating in small areas and also examines the inhibitor action in the coatings by monitoring both the cathodic and anodic corrosion activities through measuring current density maps over the selected surface of the sample. While GC-MS had been employed to determine the contents of encapsulated amines, micro-Raman spectroscopy determined the chemical composition and displayed the bands. Mechanical characterization is an essential aspect when characterizing a smart coat. It also determines functional recovery in self-healing coatings. Static fracture testing, fatigue testing, tear testing, microcapsule-induced toughening, microhardness test, scratch hardness test, pencil hardness test, and cross hatch adhesion test are among the relevant tests in use [17, 18, 22, 23].

4. Doping and heterostructures of 2D materials

There are different parameters that impact on properties of 2D materials. Substrate is one of them that is a new modification parameter in the synthesis of 2D materials. To produce any thin film, at any of its dimensions, typically needs a substrate. Combining novel substrates with 2D materials can make unusual properties.

Another effective way to modify properties is doping and functionalization with other atoms or molecules. If an even number of atoms is omitted, the structure will be more unstable and more chemically active due to hanging bonds. Therefore, reactive sites could subsequently be doped with foreign atoms to improve the specificity of the hanging bonds for binding other molecules.

It is important to be reminded that the "doping" here is different from the one generally being used in semiconductor physics. Replacement of heteroatom atoms in materials lattice can sometimes reach relatively high doping levels [24].

Magnetism properties of materials can be improved by using both electric field and electrostatic doping. Some researchers demonstrate that doping can control magnetism in both monolayer and bilayer [25].

Overall, the doping of materials like graphene could be roughly classified into two groups: electrical doping and chemical doping. Electrical doping is done by varying the gate voltages, and chemical doping is prepared using chemical methods that can be considered as an efficient method [25, 26].

Yang et al. demonstrated that a chloride molecular doping method decreases the contact resistance (R_c) in the few-layer WS_2 and MoS_2. This result makes high electron doping density; therefore, Schottky barrier width declines significantly. This doping method offers a highly practical technique to reduce the R_c in 2D materials and improve the performance in 2D nanoelectronic devices [27].

5. Coating of 2D materials using metal organic frameworks

Metal organic frameworks (MOFs) or porous coordination polymers (PCPs) as distinctive porous solids have attracted great attention. MOFs have a number of advantages rather than the other customary porous materials (e.g., porous carbon, mesoporous silica, and zeolite) such as high surface chemistry, structural flexibility, and variable pore size in the nanometer range. Therefore, inserting 2D nanomaterials and their hybrids in MOF materials can develop their potential applications [28, 29].

The encapsulation of other types of 2D nanomaterials and their related composites in MOF matrices to form 2D core-shell structures has been demonstrated by other researchers. For example, Huang et al. displayed the coating of numerous 2D materials counting MoS_2, graphene oxide (GO), and reduced graphene oxide (rGO) nanosheets as well as their hybrids with metal nanoparticles (i.e., Pt-MoS_2, Pt-GO, and Pt-rGO) with MOFs (i.e., ZIF-8). They have demonstrated the coating of various 2D nanomaterials such as MoS_2, GO, and rGO nanosheets and their hybrids with metal nanoparticles (i.e., Pt-GO, Pt-rGO, and Pt-MoS_2) with ZIF-8 that 2D materials as core and MOFs as shell are in these 2D core-shell structures. The TEM image of the Pt-rGO@ZIF-8 and Pt-rGO@ZIF-8 hybrid material are shown in **Figure 1a** and **b**, respectively [30].

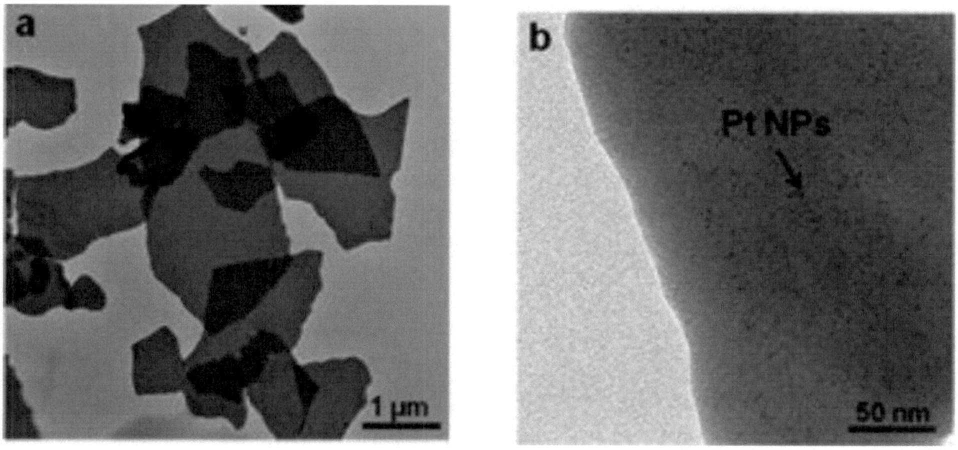

Figure 1.
(a) TEM image of Pt-rGO@ZIF-8 hybrid nanostructures. (b) Magnified TEM image of a Pt-rGO@ZIF-8 hybrid nanostructure [30].

6. Application of some thin film of 2D material

In recent years, layered transition metal dichalcogenides (TMDs) have been studied in the applications of photodetectors, photocatalysis, solar cells, optical modulators, and so on that some examples of their application will discuss in the following sections. For example, a great number of active sites could be created by the 2D-layered materials with high-specific surface areas for numerous reactions. Furthermore, nanosized 2D-layered materials could use as a support to produce different composites with a large interfacial contact. Various 2D-layered materials such as MoS_2 composed of Mo and few-layer MoS_2 nanosheet coated TiO_2 nanobelts exhibit photocatalysis activity [31].

6.1 Strong saturable absorption of 2D SnS_2 nanosheet film

SnS_2 thin film is a kind of transition metal dichalcogenide monolayers with a structure of an atomic layer of Sn between two atomic layers of S (S-Sn-S layers) as shown in **Figure 2**. The range of bandgap is from 2.1 to 2.81 eV. It has many applications such as field-effect transistors, photodetectors, and photocatalysts. The range bandgap of SnS_2 is close to the MoS_2 and WS_2. Therefore, SnS_2 thin film can have high saturable absorption property like these materials. SnS_2 can be also considered as a proper saturable absorption material with low cost. He et al. deposited SnS_2 film on transparent quartz using magnetron sputtering method by sulfuration of Sn film. The thickness of a SnS_2 thin film is measured to be 18.4 nm. The saturable absorption intensity of SnS_2 thin film was higher than other thin films that the SnS_2 thin film displayed strong saturable absorption behavior to use potential applications in mode-locked lasers for femtosecond pulse generation [32].

6.2 Water purification using 2D nanostructures

One of the global concerns is issue of the insufficient availability of drinkable water. Nowadays, pollution of water has increased. Nanoporous graphene membranes could be considered an ideal separation membrane because of its 2D nanostructures, large surface area, and transporting selected molecule properties (**Figure 3**) [33].

However, graphene is impermeable against all gases and liquids. Moreover, graphene pores were engraved with ion etching and modified with negatively charged nitrogen and fluorine and also positively charged hydrogen atoms. Therefore, pore selectivity will be improved using coupling between ions and functional groups positioned at the edge of the nanopore [34].

The passage of Li^+, Na^+, and K^+ ions is permitted by F-N-pore, and H-pore permits penetration of Cl^- and Br^- ions. Moreover, ion size has impact on the flow rate. Finally, the pore size, shape, and number of functional ligands attached can be effective on ion selectivity to enhance performance of the graphene membrane.

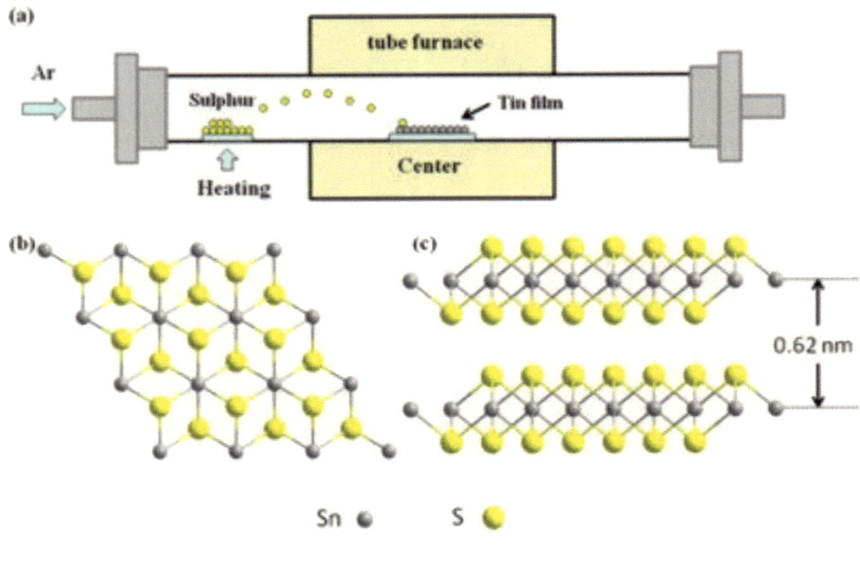

Figure 2.
(a) Schematic diagram of preparation of SnS_2 film. (b) Top view and (c) side view of the atomic structures of SnS_2 [32].

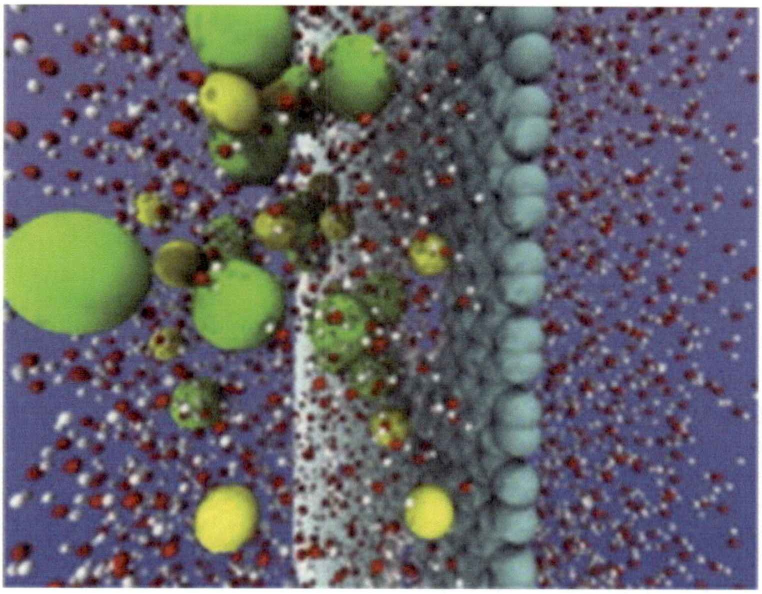

Figure 3.
Nanoporous graphene membrane for desalination [35].

Recently, use of 2D thin film materials for water treatment has attracted attention of researchers from the industry. A multilayered graphene thin film can be prepared on a support using shear force alignment to spread the viscous, uniformly arranged fluid. Multilayered graphene thin film for selective sieving can be used to charge and uncharge organic probe molecules and monovalent and divalent salts with rejection efficiency over 90%. Currently, Arvia Technology based on water and wastewater treatment uses a graphene-based thin film membrane for the treatment of toxic wastewater [35, 36].

6.3 2D materials for energy storage applications

Energy crisis is one of the critical issues in our modern society. Currently, demand for efficient, low cost, lightweight, flexible, and environmentally has increased. Therefore, use of any scale energy storage devices has raised. There are high performance energy storage devices with a high energy density at high power such as 2D materials' graphene thin film or other layered systems. Graphene has exposed excessive potential in energy storage applications as active components because of their remarkable electrochemical properties, elemental compositions, and different crystallographic structures, which can act as electrode materials for high-performance electrochemical energy storage device. Other 2D materials as layered materials of graphene analogues (GAs) refer to layered materials with a structure similar to graphene, with planar topology. This thin film can be used in energy storage. There are numerous attempts using a single-layer sheet and size to preserve their morphology to synthesize a wide range of GAs homogeneously and to control their thicknesses.

A fundamental understanding on the structures of electrodes, the electrode/electrolyte interfaces, and charge storage mechanisms plays a key role in layered material application in energy storage. Additionally, the effect of defects on the electrochemical properties of layered materials needs to be studied. The overall electrochemical performance (e.g., gravimetric energy density and power density) will improve by increasing the operating voltage and choosing an appropriate electrolyte with a high operating voltage window (organic electrolytes or ionic liquids) [37].

The lithium ion intercalation mechanisms play a key role in Li-ion batteries. **Figure 4a** shows the main power sources working for movable electronics and electric vehicles. However, lithium-ion batteries based on graphite anodes and customary cathode materials (with stable layered structures such as layered lithium metal phosphates and lithium transition metal oxides) are useful for energy storage devices with high energy density and efficiency and low cost. To develop energy storage devices with a larger capacity, higher rate ability and longer cycle life are necessary. Li-alloy-based anode materials such as silicon, tin, and tin oxide can exhibit high capacity. In recent years, extraordinary capacities in Li-ion batteries to store energy can be seen in batteries based on pure lithium metal anodes, such as nonaqueous lithium-oxygen (Li-O_2) batteries (**Figure 4b**) and lithium-sulfur (Li-S) batteries (**Figure 4c**). Li-O_2 batteries produce an insulated solid state discharge product of Li_2O_2 with high potentials. Furthermore, intermediate discharge products of lithium polysulfides were produced in Li-S batteries that are soluble and can dissolve into the electrolyte, thereby migrating to the anode side [38].

6.4 Application of 2D materials in biomedical

Recently, application of TMDCs has increased in different biomedical fields, such as drug delivery agents, therapeutics, bioimaging elements, and biosensors because of its layered structure and weak interplanar Van der Waals forces between these layers [39].

6.4.1 Photothermal therapy

Currently, there are a lot of treatments of tumors through the production of highly efficient and multifunctional nanomaterials. Photothermal therapy (PTT) has a lot of advantages such as a noninvasive, controllable, and targeted strategy to remove tumor cells. The photothermal performances can be improved using functional biomedical and bioactive nanomaterials [40]. Light is an external stimulus for cancer phototherapeutic modality. Typically, photothermal agents accumulated within tumors as internal energy absorbers can be hired by PTT to convert near-infrared (NIR) light energy into heat, producing necrosis and/or apoptosis of cancer cells. Two fundamental parameters of extinction coefficient (ε) and photothermal conversion efficiency (η) are used to determine photothermal

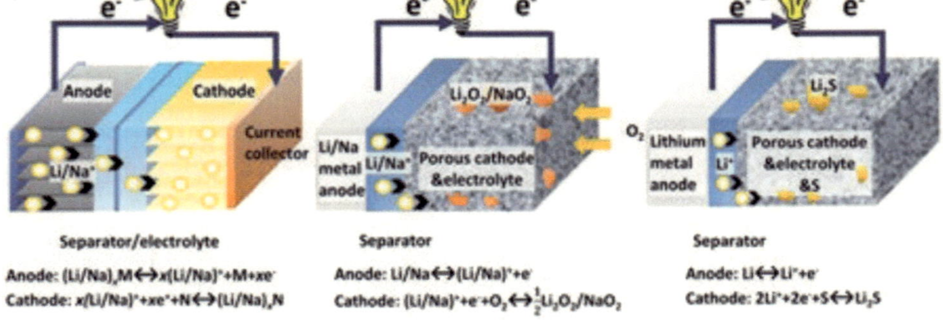

Figure 4.
Design of (a) Li/Na-ion batteries (M represents the anode material, and N represents the cathode material), (b) nonaqueous Li/Na-O_2 batteries and (c) Li-S batteries [38].

performance of a phototherapeutic agent used for photothermal conversion. 2D inorganic photothermal agents are able to increase photothermal conversion efficiency in converting the light into heat [49]. Recently, scientists have done a series of breakthroughs on exploiting ultrathin 2D transition metal carbides nanosheets (MXenes) for PTT, including effective photothermal ablation of tumors in both in vivo and in vitro as mouse model (e.g., Ti_3C_2 MXene) [41].

6.4.2 Biosensing

2D materials' thin film could also be hired as novel biosensing device for the detection of biomacromolecules and bioeffects. Recently, 2D layered molybdenum sulfide (2D MoS_2) nanosheets have displayed high potential for the development of next-generation platforms for efficient signal transduction. Opportunities to design and create highly sensitive, specific, and commercially viable sensing devices have increased through combination with DNA as a biorecognition medium and MoS_2 nanostructures. Monolayer MoS_2 nanosheets have been used as a biosensor for DNA detection, which has been confirmed to bind well with the unique transduction properties of 2D MoS_2 nanosheets. Sensors can be acted based on the principles of fluorescence, electro-chemiluminescence, and electrochemistry with many beneficial features (e.g., strong biointerfacing through various conjugation chemistries, facile sensor assembly, high stability with regard to temperature/pH, and high affinity to target) [42]. In another example, a DNA biosensor was designed based on a 2D-g-C_3N_4 nanosheet utilizing affinity changes of g-C_3N_4 to DNA probes upon their targeting of analyte and the related positron emission tomography (PET) and the fluorescence-based quenching effect. In comparison with traditional nanoparticle-based biosensors, 2D layered nanostructures have the high surface-to-volume ratio with large-area immobilization of sensing targets and the fascinating performances such as light-absorption capability and fast electron transfer fluorescence-quenching effect based on the unique physicochemical property of 2D nanosheets. Indeed, there is a continuous demand for the development of highly sensitive, selective, efficient, and cost-effective biosensing platforms [40, 42, 43].

6.4.3 Antimicrobial activity

There are different 2D layer nanostructures with potential antibacterial activities. For instance, when the Zn-Ti layered double hydroxides (LDHs) were added to bacteria suspension under visible light, the growth of microbial species such as *S. cerevisiae*, *S. aureus*, or *Escherichia coli* was strongly inhibited due to the effect of LDH unique properties and generation of ROS by Ti^{3+} under visible light. The antimicrobial behaviors of chemically exfoliated MoS_2 (ce-MoS_2) were investigated against *E. coli* and *Bacillus subtilis* bacterial culture and compared with other carbon-based nanostructures. In detail, the antibacterial activities of Ti_3C_2 MXenes were evaluated against *E. coli* and *B. subtilis* using bacterial growth curves and colony growth assays. Ti_3C_2 displays a higher antimicrobial activity toward both *E. coli* and *B. subtilis* in comparison with GO, a well-studied antimicrobial agent. The cytotoxicity of Ti_3C_2 MXenes can be measured by LDH release from the bacterial cells exposed. It can damage bacteria, and such significant change in the cell morphology/structure could be contributed by detachment of the cytoplasmic membrane from the bacterial cell wall [41].

6.4.4 Advances in biomedical applications

Graphene (G) is highly hydrophobic material when it is decorated using oxygen containing hydrophilic groups, and graphene oxide (GO) will be prepared. Due to unique surface chemistry, π-π stacking interactions and electrostatic interaction happen with other molecules in its area. Both physical and chemical bindings of drugs occur with the surface of G/GO for drug delivery applications [44]. In this section, we display these delivery mechanisms using G and GO in anticancer therapies among other fields. GO was chemically functionalized with amino groups and mixed with carboxymethyl cellulose as drug carriers, biomolecule sensors, and cellular imaging agents in anticancer therapies with a controlled and targeted release of the drug. Anticancer drugs can be released more slowly from G/GO composite supramolecular hydrogels than other hydrogel because of the higher binding attachment of hydrophobic drugs with G/GO in the gels, which is important to control release of drugs [45, 46].

6.5 Photocatalytic activity of self-assembly of layered double 2D nanoplates

Among many inorganic solids, high photocatalytic activity for visible light-induced generation of O_2 is exhibited by a few semiconducting layered double hydroxide (LDH) such as Zn-Cr-LDH. In addition, to enhance the photocatalytic efficiency of semiconducting materials, the control of band structure such as an increase in the lifetime of photogenerated holes and electrons is necessary.

The chemical exfoliation process of the host material can be used to synthesize the 2D nanoplates of the LDH material. Meanwhile, the exfoliated LDH 2D nanoplate owns positive surface charge; therefore, these can be easily assembled with negatively charged graphene nanosheets in terms of electrostatic attraction.

Strong electronic coupling between LDH and graphene species depends on the thickness of these 2D nanostructured components. Moreover, the formation of a mesoporous structure with an expanded surface area is extremely useful for the enhancement of the photocatalytic activity of component semiconductors. Gunjakar et al. synthesized the layer-by layer-ordered hybrid photocatalysts using the self-assembly of the 2D nanostructured photocatalyst with graphene nanosheets. For the first time, they studied mesoporous Zn-Cr-LDH-graphene nanohybrids with high photocatalytic activity for visible light-induced O_2 generation via the electrostatically derived self-assembly of positively charged Zn-Cr-LDH nanoplates with negatively charged graphene nanosheets [10, 47].

7. Conclusions

The thin films of 2D materials can be prepared using different methods that a single monolayer or multilayer as thin film can be fabricated by mechanical exfoliation or chemical techniques. In this chapter, we investigated the application of 2D thin films in different fields such as biomedical, energy storage, and water purification. Processing and applications of 2D layered nanostructures have developed, while there are several challenges in this field. The properties of 2D nanomaterials are highly sensitive to surface chemistry, underlying substrate, neighboring materials, and interfaces. Moreover, methods have major influence on the performance, reproducibility, and reliability of 2D nanomaterial heterostructure applications. Synthesis of 2D nanomaterials is important; therefore, careful consideration will be needed. There are also numerous important challenges to develop for commercialization.

Author details

Zahra Sayyar* and Zahra Jamshidi
Department of Chemical Engineering, Faculty of Engineering, University of Bonab, Bonab, Iran

*Address all correspondence to: z_sayyar@sut.ac.ir

References

[1] Gupta A, Sakthivel T, Seal S. Recent development in 2D materials beyond graphene. Progress in Materials Science. 2015;**73**:44-126

[2] Das S et al. All two-dimensional, flexible, transparent, and thinnest thin film transistor. Nano Letters. 2014;**14**:2861-2866

[3] Novoselov K et al. 2D materials and van der Waals heterostructures. Science. 2016;**353**:aac9439

[4] Akinwande D et al. A review on mechanics and mechanical properties of 2D materials—Graphene and beyond. Extreme Mechanics Letters. 2017;**13**:42-77

[5] Zhang H. Introduction: 2D Materials Chemistry. ACS Publications. 2018;**118**:6089-6090

[6] Dong R, Zhang T, Feng X. Interface-assisted synthesis of 2D materials: Trend and challenges. Chemical Reviews. 2018;**118**:6189-6235

[7] Smith JB, Hagaman D, Ji H-F. Growth of 2D black phosphorus film from chemical vapor deposition. Nanotechnology. 2016;**27**:215602

[8] Singh S et al. Role of 2D and 3D defects on the reduction of LaNiO 3 nanoparticles for catalysis. Scientific Reports. 2017;7:10080

[9] Li X et al. Large area hexagonal boron nitride monolayer as efficient atomically thick insulating coating against friction and oxidation. Nanotechnology. 2014;**25**:105701

[10] Gunjakar JL et al. Self-assembly of layered double hydroxide 2D nanoplates with graphene nanosheets: an effective way to improve the photocatalytic activity of 2D nanostructured materials for visible light-induced O 2 generation. Energy & Environmental Science. 2013;**6**:1008-1017

[11] Samorì P, Palermo V, Feng X. Chemical approaches to 2D materials. Advanced Materials. 2016;**28**:6027-6029

[12] Eklund P, Rosen J, and Persson POÅ. Layered ternary Mn phases and their 2D derivative MXene: An overview from a thin-film perspective. Applied Physics. 2017;**50**:113001-113015

[13] Eklund P, Rosen J, Persson POÅ. Layered ternary M n+ 1AX n phases and their 2D derivative MXene: an overview from a thin-film perspective. Journal of Physics D: Applied Physics. 2017;**50**:113001

[14] Perea-López N et al. CVD-grown monolayered MoS2 as an effective photosensor operating at low-voltage. 2D Materials. 2014;**1**:011004

[15] Lin Z et al. 2D materials advances: from large scale synthesis and controlled heterostructures to improved characterization techniques, defects and applications. 2D Materials. 2016;**3**:042001

[16] Whelan PR et al. Raman spectral indicators of catalyst decoupling for transfer of CVD grown 2D materials. Carbon. 2017;**117**:75-81

[17] Ulaeto SB et al. Developments in smart anticorrosive coatings with multifunctional characteristics. Progress in Organic Coatings. 2017;**111**:294-314

[18] Nguyen-Tri P et al. Nanocomposite coatings: Preparation, characterization, properties, and applications. International Journal of Corrosion. 2018;**2018**:474-501

[19] Zahiri SH, Yang W, Jahedi M. Characterization of cold spray titanium supersonic jet. Journal of Thermal Spray Technology. 2009;**18**:110-117

[20] Gabard M et al. Novel method based on spin-coating for the preparation of 2D and 3D Si-based anodes for lithium ion batteries. ChemEngineering. 2017;**1**:5

[21] Yan H et al. Fabrication of 2D and 3D ordered porous ZnO films using 3D opal templates by electrodeposition. Electrochemistry Communications. 2005;**7**:1117-1121

[22] Pingale AD, Belgamwar SU, Rathore JS. Synthesis and characterization of Cu–Ni/Gr nanocomposite coatings by electro-co-deposition method: effect of current density. Bulletin of Materials Science. 2020;**43**:66

[23] Sayyar Z, Babaluo AA, Shahrouzi JR. Kinetic study of formic acid degradation by Fe3+ doped TiO2 self-cleaning nanostructure surfaces prepared by cold spray. Applied Surface Science. 2015;**335**:1-10

[24] Terrones H et al. The role of defects and doping in 2D graphene sheets and 1D nanoribbons. Reports on Progress in Physics. 2012;**75**:062501

[25] Jiang S et al. Controlling magnetism in 2D CrI 3 by electrostatic doping. Nature Nanotechnology. 2018;**13**:549

[26] Han Z. Controlling magnetism in 2D CrI 3 by electrostatic doping. Journal of Semiconductors. 2019;**40**:080204-0802042

[27] Yang L et al. Chloride molecular doping technique on 2D materials: WS2 and MoS2. Nano Letters. 2014;**14**:6275-6280

[28] Cao X et al. Metal oxide-coated three-dimensional graphene prepared by the use of metal–organic frameworks as precursors. Angewandte Chemie International Edition. 2014;**53**:1404-1409

[29] He L et al. Core–ahell noble-metal@ metal-organic-framework nanoparticles with highly selective sensing property. Angewandte Chemie International Edition. 2013;**52**:3741-3745

[30] Huang X et al. Coating two-dimensional nanomaterials with metal–organic frameworks. ACS Nano. 2014;**8**:8695-8701

[31] Weigelt S et al. Surface synthesis of 2D branched polymer nanostructures. Angewandte Chemie International Edition. 2008;**47**:4406-4410

[32] He J et al. Preparation of vertically aligned two-dimensional SnS2 nanosheet film with strong saturable absorption to femtosecond laser. Journal of Physics D: Applied Physics. 2019;**52**:165101

[33] Liu H, Wang H, Zhang X. Facile fabrication of freestanding ultrathin reduced graphene oxide membranes for water purification. Advanced Materials. 2015;**27**:249-254

[34] Zhao Z et al. One pot synthesis of tunable Fe 3 O 4–MnO 2 core–shell nanoplates and their applications for water purification. Journal of Materials Chemistry. 2012;**22**:9052-9057

[35] Dervin S, Dionysiou DD, Pillai SC. 2D nanostructures for water purification: Graphene and beyond. Nanoscale. 2016;**8**:15115-15131

[36] Akbari A et al. Large-area graphene-based nanofiltration membranes by shear alignment of discotic nematic liquid crystals of graphene oxide. Nature Communications. 2016;**7**:10891

[37] Zhang X et al. 2D materials beyond graphene for high-performance energy storage applications. Advanced Energy Materials. 2016;**6**:1600671

[38] Shi L, Zhao T. Recent advances in inorganic 2D materials and their applications in lithium and sodium

batteries. Journal of Materials Chemistry A. 2017;**5**:3735-3758

[39] Li Z, Wong SL. Functionalization of 2D transition metal dichalcogenides for biomedical applications. Materials Science and Engineering: C. 2017;**70**:1095-1106

[40] Wei W et al. Biocompatible and bioactive engineered nanomaterials for targeted tumor photothermal therapy: A review. Materials Science and Engineering: C. 2019;**104**:1-21

[41] Lin H, Chen Y, Shi J. Insights into 2D MXenes for versatile biomedical applications: Current advances and challenges ahead. Advanced Science. 2018;**5**:1800518

[42] Kukkar M et al. A comprehensive review on nano-molybdenum disulfide/DNA interfaces as emerging biosensing platforms. Biosensors and Bioelectronics. 2018;**107**:244-258

[43] Kannan PK et al. Chemical and biosensors based on graphene analogue 2D materials and their hybrids. Encyclopedia of Nanoscience and Nanotechnology. 2018;**27**:63-87

[44] Tadyszak K, Wychowaniec J, Litowczenko J. Biomedical applications of graphene-based structures. Nanomaterials. 2018;**8**:944

[45] Erol O et al. Recent advances in bioactive 1D and 2D carbon nanomaterials for biomedical applications. Nanomedicine: Nanotechnology, Biology and Medicine. 2018;**14**:2433-2454

[46] Hempel M et al. A novel class of strain gauges based on layered percolative films of 2D materials. Nano Letters. 2012;**12**:5714-5718

[47] Teramura K et al. Photocatalytic conversion of CO2 in water over layered double hydroxides. Angewandte Chemie International Edition. 2012;**51**:8008-8011

Chapter 2

Thermal and Mechanical Properties of Polypropylene Polymer Nanocomposites Infused with Sonochemically Coated SiC/SiO_2 Nanoparticles

Vijaya Rangari and James Davis

Abstract

This chapter describes the coating of silicon carbide nanoparticles on different types of silicon dioxide that varied in size and shape using sonochemical method. These composite particles were further infused into polypropylene polymer to increase its thermal and mechanical properties for various applications. A two-step process was used to fabricate SiC/SiO_2/polypropylene nanocomposites. In the first step, SiC nanoparticles were coated onto four different types of SiO_2 nanoparticles. The coated nanoparticles were then characterized using a high resolution transmission electron microscope (TEM), X-ray diffraction (XRD) determined the morphology and crystalline structure, and X-ray photoelectron spectroscopy (XPS). These results showed that the nanoparticles were crystalline, spherical in shape, and were uniformly coated. In the second step, nanocomposite samples were extruded using a Wayne Yellow Label Top single screw extruder. The as prepared nanocomposite samples were then characterized for their thermal and mechanical properties. These properties show increase in their flexural strength and thermal degradation. These results show increase in mechanical properties. The importance of this work lies in the simple sonochemical synthesis of SiC/SiO_2 hybrid nanomaterials and their filler applications in polypropylene polymer nanocomposites which are widely used for various application including automotive and electronic industries.

Keywords: polypropylene, melt extrusion, coating, ultrasound, mechanical properties

1. Introduction

In recent years, there have been rapid growth predictions for polypropylene-based composites, due to their applications in many fields such as automotive, home appliances, and construction industries [1–5]. Since the last decade, several nanoparticles have been incorporated into polypropylene and other thermoplastics as reinforcements, to produce stronger and lighter polymer composites [5–9]. There are many different thermoplastic polymers available on the market today such as polypropylene, polyethylene, etc. Amongst these, polypropylene (PP) is one of the most widely used polymers due to the fact that it is inexpensive and offers attractive properties (rigidity, light weight, thermal and chemical stability, etc.); however, it has the disadvantage of being relatively brittle at room temperature and exhibiting poor resistance to crack propagation. Many investigations have been reported on fracture properties of polypropylene. It has thus been shown that fracture toughness of PP increases with increasing molecular weight and decreasing crystalline percentage. The temperature at brittle-ductile transition of PP has also been reported to increase with increasing crystallinity [10–13]. Most properties of polymers can be altered through the addition of fillers. Recent polymer nanocomposites have been an area of intense research because of their potential applications as multifunctional and high-performance materials. These nanocomposites are fabricated by simply adding small amounts of nanoparticles in the polymer matrices. The general class of nanocomposite is the mixture of organic and inorganic materials. Significant effort is focused on the ability to obtain control of the nanoscale structures via innovative synthetic approaches. This is a rapid growing area of research and is expanding as it generates many new exciting materials with various properties. These properties of nanocomposite materials depend not only on the properties but also on their morphology and interfacial characteristics. The nanoparticles significantly enhance the mechanical properties, thermal stability, reduced gas permeability, and other physicochemical properties [14–19] of the neat polymer. To change the properties of these composites, relatively small amounts of the nanoparticle are needed, more often the percentage by weight (0.5–5%) which results from the particles having incredibly high surface to volume ratios.

Since the discovery of nanoparticles, researchers are aiming to achieve a uniform particles-polymer interaction rather than the polymer-polymer or particle-particle interaction to obtain the maximum benefit. Wetzel et al. [20] incorporated various amounts of Al_2O_3 nanoparticles into a polymer matrix and studied the influence of nanoparticles on the flexural strength, impact energy, dynamic mechanical thermal properties, and block-on-ring wear behavior, and their results show the improvement in nanocomposites as compared to the neat polymer. Zheng et al. [21] employed high frequency ultrasonic and mechanical procedures to disperse spherical SiO_2 in epoxy resin. The results indicated that with the addition of 3 wt% of SiO_2 into the resin, the tensile strength improved by 114%, the stiffness improved by 13%, and the impact strength improved by 56%. Wang's et al. [22] reported that the dispersion of TiO_2 nanoparticles was coated by an acrylic acid (AA)-plasma-polymers in a glycol solution, and results indicated that the dispersion of TiO_2 nanoparticles were greatly improved after AA-plasma-polymer coating. They noted that the surface energy played a vital role in the dispersion behaviors of TiO_2 nanoparticles. It is an effective way to improve the dispersion of nanoparticles by changing their surface energy by plasma polymer coating. Dai et al. [23] synthesized β-SiC nanorods using the reaction of carbon nanotubes by using a mixture of Si and I_2 or through the use of a volatile SiO_2. In this study, there was a large quantity of β-SiC nanorods prepared which was wrapped with uniform amorphous SiO_2 layers on the outside surface. The formation of β-SiC nanorod was based on the carbothermal reduction of silica xerogels with carbon nanoparticles embedded in the network, and the formation of SiO_2 layer was from the reaction of decomposed SiO and O_2.

Extrusion is one of the most widely used processing techniques for fabrication of thermoplastic polymers. It can be used as a "stand alone" machine to directly shape parts or it can also be used as a melting device that is coupled with other secondary shaping devices. The two main advantages of this process over other manufacturing processes are its ability to create very complex cross-sections and work materials that are brittle, because the material only encounters compressive and shear stresses. It also forms finished parts with an excellent surface finish [24, 25]. Extrusion is also used in the dispersion of nanoparticles to prepare polymer nanocomposites. Mohammed et al. [26] investigated the effect of SiO_2 infused into nylon-6 at wt% of 0, 1, and 2, respectively, and extruded into filaments. Experimental results showed that by infusing nanoparticles into nylon -6 and extruding it increased its tensile modulus, yield strength, hardening modulus, and ultimate tensile strength. There was a 45% enhancement in tensile modulus and a 26% enhancement in ultimate tensile strength observed in the 2 wt% system while compared to the neat nylon-6. TGA results indicated more thermal stability found in the nanophased infused systems, while the DSC studies indicated a more moderate increase in T_g. Liu et al. [27] presented the effects of incorporated montmorillonite (MMT) on a surface and the bulk mechanical properties of as-synthesized polyamide-6/montmorillonite (PA6/MMT) composites that were prepared using the twin-screw extruder mixing technique. Russo et al. [28] studied the effect of multiple extrusions on nanostructure and properties of nylon 6 nanocomposites. These researchers produced nanocomposites at different silicate loadings by melt compounding and submitted to further reprocessing by using single and twin screw extruders. Rheological, morphological, and mechanical analyses were carried out on as-produced and reprocessed samples to explore the influence of the number and the type of extrusion cycles on silicate nanodispersion. The results obtained displayed that the reprocessing by single screw extruder can modify the initial morphology since the re-agglomeration of the silicate layers can occur. However, a better nanodispersion was observed in the hybrids reprocessed by twin screw extruder, which was a result of the additional mechanical stresses able to realizing a dispersive mixing that contributes to avoid re-agglomeration phenomena. The high shear stresses produced with twin screw geometry determined also a significant degradation of neat matrix, principally based on chain scission mechanism. All as-produced and reprocessed hybrids showed a substantial enhancement in tensile modulus with the adding of silicate. However, the entity of performance enhancements displayed by the reprocessed hybrids was found to be highly dependent on the degradation of both organoclay and polymer matrix as well as the silicate amount, the number, and the type reprocessing. Investigations were made by Yong et al. [29] on the effects of the coating amount of surfactant and the particle concentration on the impact strength of polypropylene (PP)/$CaCO_3$ nanocomposites. These nanocomposites prepared with monolayer-coated $CaCO_3$ nanoparticles had the best mechanical properties, including Young's modulus, tensile yield stress, and impact strength because of the good dispersion of the nanoparticles in the polymer matrix, which in turn allowed them to study the effects of particle concentration on the impact strength of the nanocomposites. H-PP and E-PP, which were the low and high molecular weight PPs, respectively, were used as polymer matrices. Critical particle concentrations of 10 and 25 wt% corresponding to an abrupt increase in the impact toughness were determined for the E-PP and H-PP nanocomposites, respectively. Manolis et al. [30] observed a series of experiments which were performed on both simply supported and on-grade circular slab specimens, reinforced with different volumes of fibrillated polypropylene fibers in order to gauge its influence on the slab's impact resistance and natural frequency. Literature survey clearly suggests that there have been studies on polymer systems infused with nanofillers, but there are no reports on combination of silicon carbide, silicon dioxide coatings infused into polypropylene. Herein, this study we report the effects of reinforcement of silicon dioxide coated silicon carbide nanoparticles in polypropylene thermal and mechanical properties.

2. Experimental

2.1 Materials

The nanoparticles used in these experiments were spherical silicon carbide, ~30 nm in diameter. The β-silicon carbide nanoparticles were purchased from MTI Corporation, Richmond CA, USA. These nanoparticles were then coated with four different types of silicon dioxide nanoparticles, colloidal (Nissan chemicals), 80 nm, and areosil (Degussa), and needle shaped SiO_2 nanoparticles (Nissan chemicals). The colloidal silica solution MP-1040 was procured from Nissan Chemical Corporation. This filler is composed of 40% nanoparticles that are spherical in shape and approximately 200 nm in diameter dispersed in 60% of water. The 80 nm silica was procured from MTI Corporation Richmond CA, USA. These particles are spherical in shape having an 80 nm diameter. Areosil silica is also spherical in shape with a smaller diameter of approximately 5–15 nm and was also procured from Degussa. The other silica used was the needle shaped silica solution MP-1040 which was procured from Nissan Chemical Corporation. This filler is also composed of 40% nanoparticles and 60% water; however, these nanoparticles are irregular in shape.

2.1.1 Polypropylene

The matrix material used in this experiment is polypropylene procured from Chem Point. The polymer molecular weight ranged from 40,000 to 100,000, its powder diameter is approximately 20 μm, and the density of polypropylene is between 0.910 and 0.928 gr/ml. The melting point of polypropylene is 160°C with >50 isotactic crystalline structure. This type of high melting polymer can be molded or machined into structural components [31].

2.2 Synthesis of SiC/SiO_2 hybrid nanoparticles

In the first step, 250 mg of β-SiC was weighed and mixed along with 500 mg of powdered SiO_2, 250 mg of pluronic F-127 and 60 ml of ethylene glycol. The reaction mixture was irradiated using high intensity ultrasound for 3 hours at 5°C under argon gas flow. This process was done for each of the two powdered types of SiO_2; however for the solution types, when added to SiC nanoparticles, it was done using a weight ratio of 10:1. Upon completion of the sonication process, the silicon carbide coated silica nanoparticles were collected using a centrifuge at 12,000 rpm at 5°C for 30 min. The coated particles were washed 5 times, the first four times using distilled water and the last time with ethanol. The final step was to allow the sample to dry under vacuum for 24 hours at room temperature. Once dried, the coated nanoparticles were added to polypropylene at a ratio of 1:10 and then mixed using a Thinky mixer. The materials were placed in a special container and mixed under vacuum at 1200 rpm for 10 min. This step was done three times to ensure proper distribution of particles, before being extruded. This non-contact mixing method [32, 33] proves to be very efficient at dispersing nanofillers in polymer matrices without having any cross-contamination. This non-contact method works by independently revolving and rotating a container at various speeds in a planetary manner. As the container rotates, it experiences a number of forces that results in the material inside of the container being mixed. Centrifugal force presses the material outwards to the inner wall of the container, while other forces cause the material to move in a spiraling downwards direction along the slope of the container wall. The material then moves back to the center of the container and up to the top repeating the process and in turn mixing the sample. As the material circulates from top to bottom, it generates a kneading effect forcing trapped air out of the material.

2.3 Extrusion

In the second phase, nanocomposites of polypropylene, silicon carbide/silicon dioxides were then extruded using a Wayne Yellow Label Top single screw extruder. The extruder has a 19 mm diameter screw, which is driven by a 2HP motor complete with a toothed timing belt for smooth speed reversal. Present are five thermostatically controlled heating zones that were used to melt the mixture before extrusion. Three of the zones were located inside the barrel and the final two were in the die zone set at respective temperatures of 340, 344, 350, 354, and 350°F. As the sample material passes through each zone, they are disintegrated into several branches and then combined again, which ensures proper distributive mixing of the nanoparticles and polypropylene. As the dispersed nanoparticles containing liquid polypropylene passes through the 10 cm long steel tube, it arrives to the die plate. As the sample passes through the extruder, it enters and completely fills the die, which in turn forms the flexure sample. The process schematic is shown in **Figure 1**.

2.4 Characterization

2.4.1 Transmission electron microscopy (TEM)

JOEL-2010, High Resolution (Japan), Transmission Electron Microscopy (HRTEM), is used to study the size, shape, and extent of coating of SiC onto SiO_2 nanoparticles. The samples were prepared by dispersing the nanoparticles in ethanol and placing a drop of the solution on a copper grid.

2.4.2 X-ray diffraction (XRD)

Rigaku D/MAX 2100 X-Ray Diffractometer (Japan) was used to study the structural characteristics. The samples were prepared for XRD by uniformly spreading the nanoparticles on a quartz sample holder. The test samples were conducted from 0 to 80° of two theta at room temperature. Characteristics XRD diffraction patterns were collected and matched with established data from PDF files. X-ray diffraction helps to determine the crystallographic configuration of the coated nanoparticles and its nanocomposites.

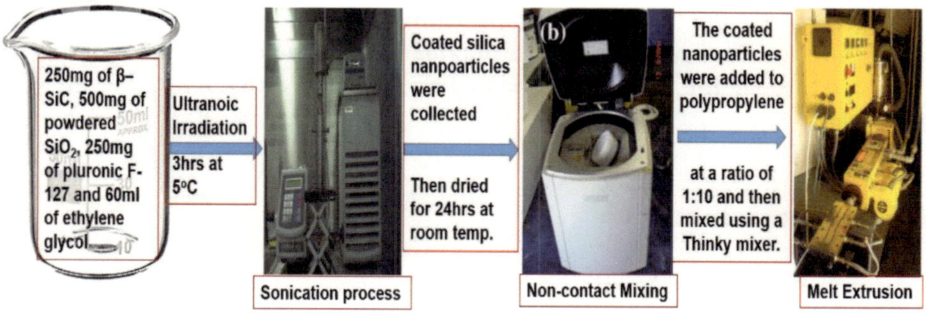

Figure 1.
Fabrication of polypropylene/SiC/SiO_2 nanocomposite.

2.4.3 X-ray photo electron spectroscopy (XPS)

This surface analysis technique (2–10 nm analysis depth typically) XPS can identify elemental composition, oxidation state, and chemical bonding like covalent bonds (oxides, nitrides, etc.) by evaluating binding energy shifts. This test was done at Cornell University Dr. Jonathan Shu's lab. The pressure in the analyzing chamber was typically 1.3×10^7 Pa under operating conditions. Photoelectrons were collected at approximately 45° from the specimen surface normal. The X-rays created photoemission electrons, and the signature of the photoemissions identified each element present in the samples.

2.4.4 Thermogravimetric analysis (TGA)

Thermogravimetric analysis was carried out using a Mettler Toledo (USA) TGA apparatus to find out the thermal stability of the test sample. This test is most commonly used to determine polymer decomposition temperatures, residual solvent levels, the amount of moisture content absorbed, and the amount of inorganic filler in polymer or composite material compositions. The samples were prepared by cutting the samples into small pieces, then placing 5 mg of the material in a platinum sample pan, and heating at a constant heating rate of 10°C/min from room temperature to 900°C under nitrogen atmosphere of 40 ml/min. The initial sample weight is measured at room temperature, and as heat is constantly being applied, the machine monitors the change in sample weight as a function of temperature. This result in weight loss versus temperature graphs being plotted from the data is obtained. In this case, the decomposition temperature is considered at 50% weight loss of the material.

2.4.5 Differential scanning calorimetry (DSC)

Differential scanning calorimetry (DSC) tests were carried out using a Mettler Toledo822^e (USA) from 30 to 300°C at a heating rate of 5°C/min under a nitrogen atmosphere then cooled from 300 to 30°C at the same rate. These results were used in measuring the melting temperature and the crystalline temperature.

2.4.6 Flexural analysis

Three-point bending flexural tests were conducted using a Zwick Roell MTS (Germany) to evaluate the samples flexural properties. Ten samples were tested using ASTM D790-03 testing standards which are of nominal size 100 mm (length) × 12.5 mm (width) × 5 mm (thick) and a span/thickness ratio maintained at 16:1. The test were conducted at room temperature at a constant crosshead speed of 2.0 mm/min. Stress-strain plots were obtained as the load displacement data were recorded through the data acquisition system equipped with the Zwick Roell MTS machine.

3. Results and discussion

3.1 X-ray diffraction (XRD)

XRD patterns were collected using a diffractometer operating in the Bragg-Brentano geometry utilizing Cu Kα radiation. The X-ray tube was operated at 40 kV and 30 mA. A 2θ scan range from 30 to 80° in steps of 2^0/min was recorded.

The SiC coated onto various SiO_2 samples were analyzed to investigate the effects of ultrasonic irradiation on the nanoparticles. The XRD spectra of pure SiC can be seen below in **Figure 2(a)**. Through the use of the Jade 6.0 software, the peaks appearing at 2θ of 35.6, 41.5, 60.0, 72.0, and 75.5° indexed as (111), (200), (220), (311), and (222), designates the crystal spherical structured Moissanite-3C SiC (JCPDS 29-1129). The XRD of silicon carbide coating onto silicon dioxide was shown in **Figure 2(b)–(e)**, and it is evident that all of the SiC particles are crystalline in nature and all assigned to β-SiC. The background wide peak pattern is the characteristic of an amorphous SiO_2 phase. In the sample, the XRD data indicate the presence of only silicon carbide and amorphous silicon dioxide. The plot has characteristic peak intensities of FWHM, and d-spacing values which can be found in **Table 1**. A sharp peak around 35.6° corresponds to a spacing of 2.5189 A, which is attributed to the β-SiC (111) diffraction. There was an amorphous background present around 22° which is the result of the silicon dioxide presence. Deduced from the FWHM of β-SiC (111) diffraction peak. It is evident that, compared to pure SiC, the coated SiC FWHM peaks decreased significantly. In this case, the coated nanoparticles FWHM decreased compared to the neat, which results in coating of amorphous silica. The crystallite sizes of the particles were determined using Debye-Scherrer formula and FWHM method from the 100% peak of the pattern.

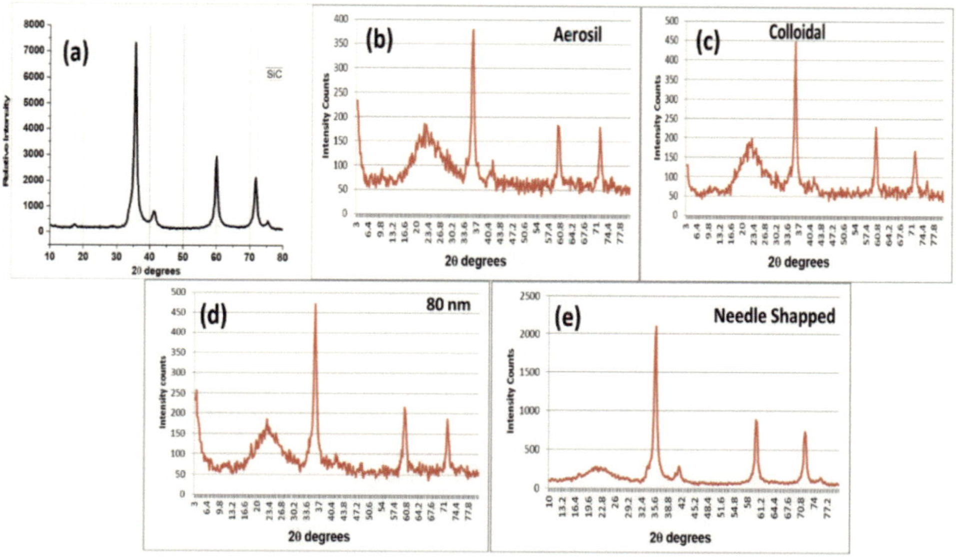

Figure 2.
XRD patterns of SiC and coated nanoparticles.

Sample	2θ at 100% intensity	FWHM(°)	FWHM at 35.6°
SiC	35.612	0.673	0.673
Colloidal	35.746	0.389	0.612
Areosil	35.813	0.397	0.604
80 nm	35.694	0.343	0.674
Needle shaped	35.865	0.368	0.652

Table 1.
XRD analysis of SiC and coated SiC, FWHM.

3.2 Transmission electron microscopy (TEM)

The TEM micrograph in **Figure 3** shows (a) as received SiC, (b) SiC coating on colloidal silica, (c) aerosol silica coated on SiC, (d) 80 nm silica coated on SiC, and finally (e) needle shaped silica coated on SiC. **Figure 3(a)** shows the SiC particles are spherical and particle sizes are about 30–50 nm. **Figure 3(b)** represents the coating of SiC on colloidal silica where the particles of silica are about 200 nm. The SiC particles are uniformly coated on silica spheres. **Figure 3(c)** shows that the particles of aerosol silica is coated on SiC nanoparticles. Since the aerosil silica is amorphous and also the particle sizes are about 5 nm, it is seen on the surface of SiC. In TEM **Figure 3(d)** and **(e)** shows the SiC particles and it was very hard to see the silica because of the silica is highly transparent to the electronic bean and difficult image.

3.3 X-ray photoelectron spectroscopy

XPS is a surface sensitive spectroscopic tool that provides information about chemical states and concentration of elements present in a sample with the exceptions of H and He. Each characteristic peak corresponds to the electron configuration of the electrons within the atoms, e.g., 1s, 2s, 2p, 3s, etc. This method uses soft X-rays to eject electrons from inner-shell orbital. The kinetic energy, E_k, of these photoelectrons is determined by the energy of the X-ray radiation, $h\nu$, and the electron binding energy, E_b, as given by:

$$E_k = h\nu - E_b - \emptyset \quad (1)$$

$$h - \text{Plank's constant } (6.62\text{x}10-34 \text{ J/s}).$$

Using this technique, the surface of SiC and SiO_2 were characterized from the surface to approximately 10 nm in depth. Each element shows a characteristic set of peaks in the photoelectron spectrum at kinetic energies of the elemental

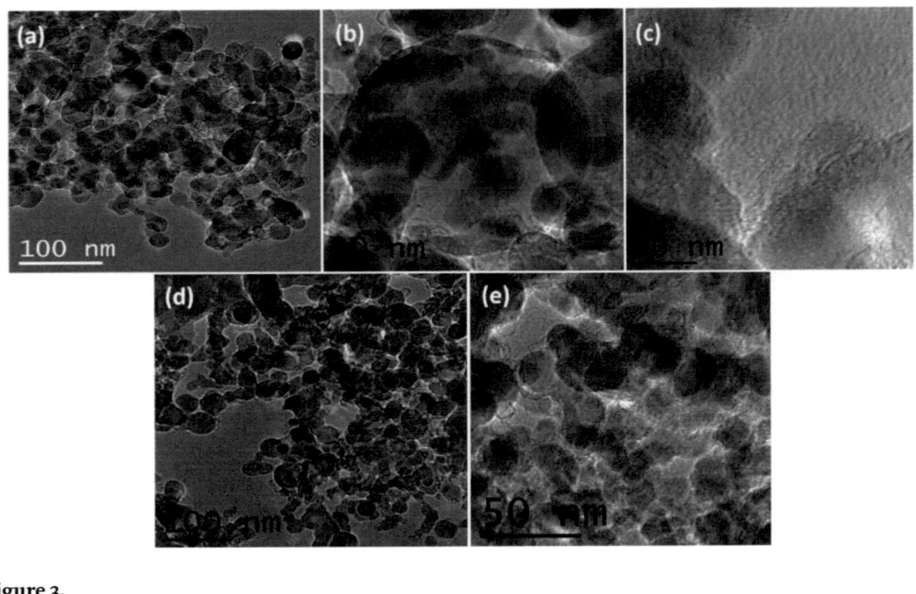

Figure 3.
TEM images of (a) SiC, (b) SiC coated SiO_2, (c) SiC coated areosil, (d) SiC coated needle SiO_2, and (e) SiC coated 80 nm SiO_2.

composition of the surface region determined by the photon energy and the respective binding energies. Elements exhibit binding energy peaks whose relative position depends on the electronegativity of their surrounding atomic neighbors. In the case of a phase mixture where more than one bonding state exists, a particular element is expected to have various binding energies, caused by different coordination/neighbor. The intensity of the peaks is related to the concentration of the element within the sampled region.

As shown in **Figure 4**, it displays the XPS spectra of SiC/SiO_2 nanomaterials. The most distinctive peaks present are the oxygen, carbon, and silicon peaks, respectively. Using the CasaXPS software for XPS analysis, a detailed spectrum of Si 2p, C 1s and O 1s of the chemical state of each element on the surface of the nanoparticles were identified and analyzed. Analysis of such element regions were done to confirm their attachment to the surface of the silicon dioxides. Silicon (Si 2p), carbon (C 1s), and oxygen (O 1s) elements can be found at binding energies of 102.8, 284.6, and 532.0 electron volts (eV), respectively. Also in **Figure 4** above, one can take a look at the as received SiC which clearly shows a Si peak, but also shows where adsorbed contaminations like oxygen and carbon were found on the top layer within about 6 nm in thickness. This observation is consistent with other researchers [34, 35] and can be a result of the absorption of atmospheric oxygen. **Figure 5(a)** shows the high resolution scan of the deconvoluted XPS spectra of neat SiC for Si 2p peak and C 1s on the sub-surface layer. The spectra were Gaussian fitted based on the published data of the binding energies. The spectra emission line at binding energy (BE) of 100.47 eV is associated with the SiC bonding. In addition to this, two more fitted chemical shifts were observed at BE of 98.4 and 102.9 eV, which were identified as Si° and SiO_2, respectively, for the spectra of SiC for Si 2p peak.

For the spectra of SiC, the C 1s spectrum was deconvoluted into three peaks (**Figure 5(b)** of BE 285.5, 282.7, and 289.06eV. The peak at 285.5 eV signified the presence of carbon in the form of hydrocarbon, while the peak at 282.7 eV is due to the binding energy of carbon in the form of SiC, and in the form of carbon bonded to oxygen, this was observed at 289.06eV. The binding energy for oxygen or O 1s was found to be 532.6 eV and can be associated either with O_2-Si or C-O bonding which can be a result of surface contamination. From the surface chemical composition, it can be seen that the ratio of Si/C was found to be 0.29, which is much lower than the theoretical value for pure SiC of 0.5 which clearly indicates surface contamination.

The coated hybrid nanocomposites of colloidal, areosil, 80 nm, and needle shape nanocomposites were also investigated using XPS for comparison with the neat sample. The chemical composition and binding energies were tabulated and

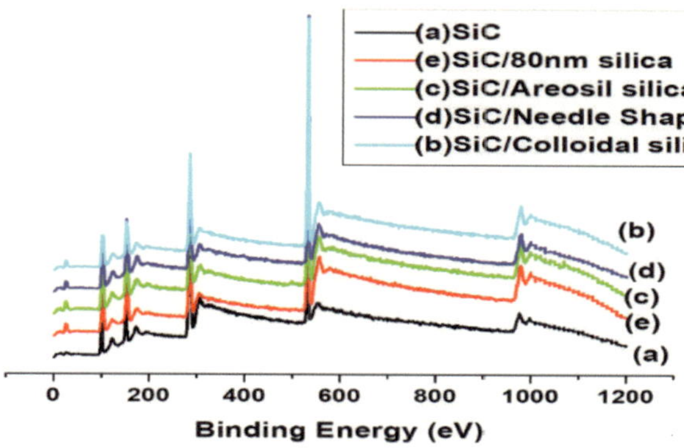

Figure 4.
XPS spectra for neat SiC and coated nanoparticles.

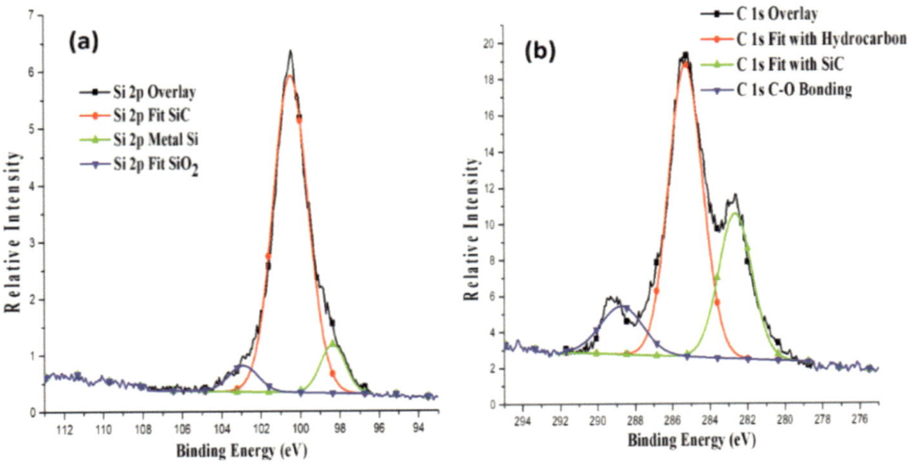

Figure 5.
Deconvoluted XPS spectra for SiC for (a) Si 2p peak (b) C 1s peak.

presented in **Table 2**. The O/Si ratios from the table obviously doubled which indicated that surface contaminant oxygen was increased when SiC was sonicated with the four different types of SiO_2. The concentration of C in the near surface decreases from 65.09% for neat SiC to an average of 43.26% for sonicated SiC in SiO_2. This decrease is may be removal of carbon in the form of CO_2. However, it can be seen here that, the as received SiC absorb oxygen which is enough to impart strong chemical double bonding in the system.

Figures 6–9 depict the deconvoluted XPS spectra for peaks Si 2p, C 1s, and O 1s of the coated samples of colloidal, areosil, needle shaped and 80 nm nanoparticles. For each of the nanoparticles samples made using high intensity ultrasound irradiation, there were noticeable chemical shifts in peak observed indicating the coating of SiC onto SiO_2 nanoparticles that took place. This was further confirmed from the elemental concentration at the surface. These spectra could be separated into two or three peaks based on the assumption that each peak consists of the Gaussian/Lorentzian sum function. The Si(2p) peak was separated into the Si–Si, Si–C, and SiO_x peaks, and the C(1s) peak was decomposed into the C–Si, C–C, C–O–H, and C=O peaks [34–39]. The energy positions of these peaks are listed in **Table 3**. The separation of the elements is not sufficient to precisely obtain the fraction of the bond. Therefore, as a rough estimation, the fraction of the bond was obtained from the ratio of the area of each peak [40]. The FWHM of Si–Si and Si–C are 2.13 and 1.38 eV, respectively. The FWHM of C–C and C=C are 2.1 and 1.2 eV, respectively. The ratio between C–C and C=C bond concentration is 10.1.

Samples	Composition (atomic %) at the surface of nanoparticles			Atomic ratios of O/Si
	Si	O	C	
As received SiC	19.05	15.86	65.09	0.83
Colloidal	20.91	33.64	45.45	1.60
Areosil	23.59	34.51	41.35	1.46
80 nm	19.82	37.28	42.90	1.88
Needle shaped	25.66	30.97	43.37	1.20

Table 2.
Surface chemical composition of neat SiC and coated nanoparticles.

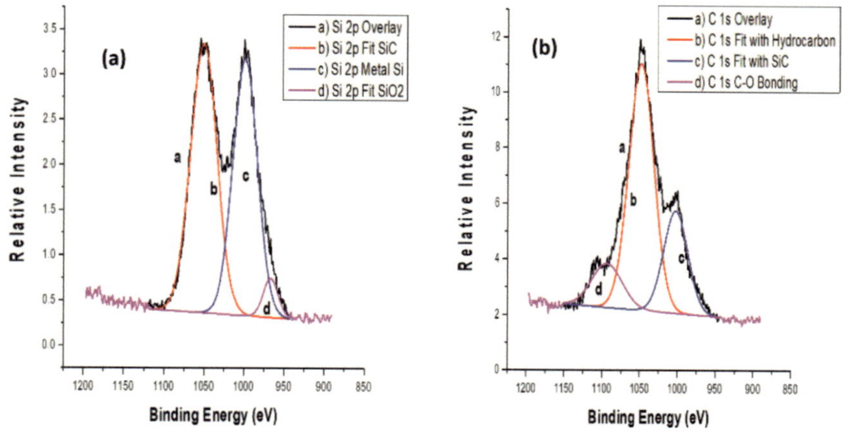

Figure 6.
Deconvoluted XPS spectrum of SiC/SiO_2 (colloidal) for Si 2p peak and C 1s.

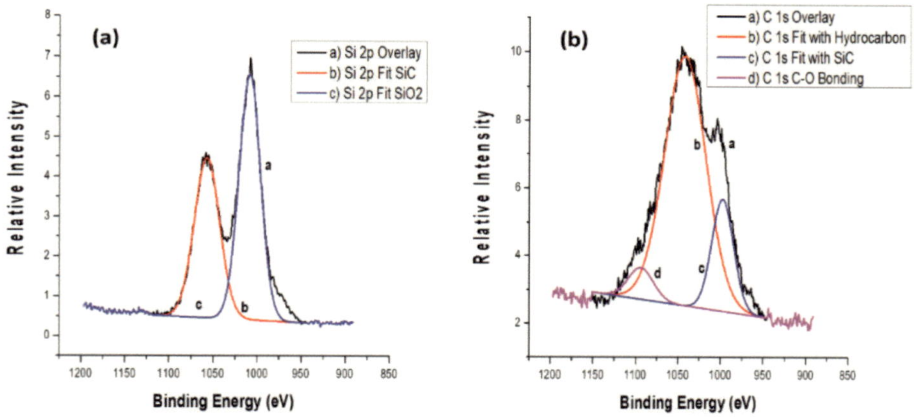

Figure 7.
Deconvoluted XPS spectrum of SiC/SiO_2 (Areosil) for (a) Si 2p peak and (b) C 1s.

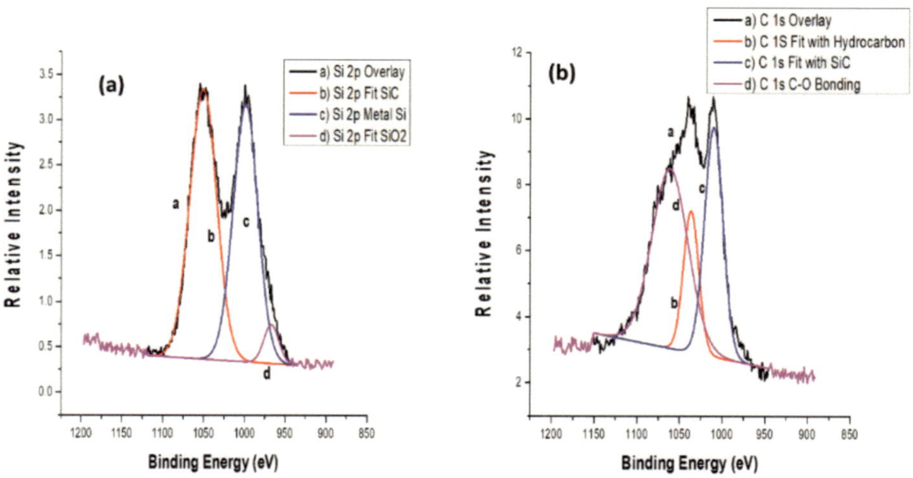

Figure 8.
Deconvoluted XPS spectrum of SiC/SiO_2 (needle shape) for (a) Si 2p peak and (b) C 1s.

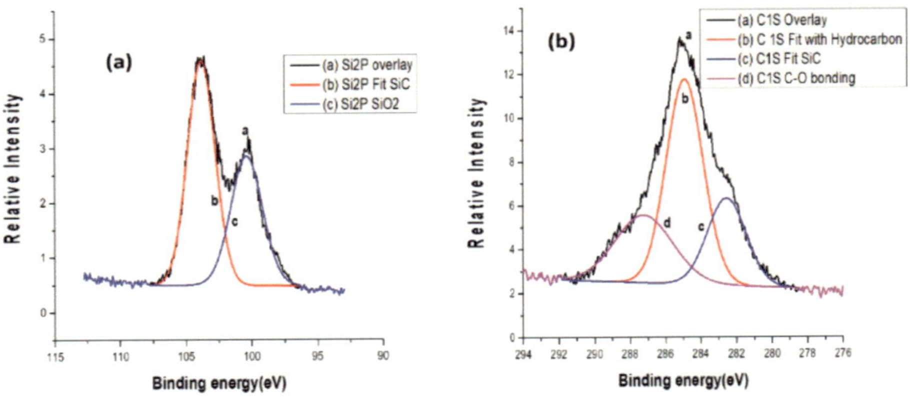

Figure 9.
Deconvoluted XPS spectra of SiC/SiO$_2$ (80 nm) for (a) Si 2p and (b) C 1s.

Bond	Binding energy (eV)
Si–Si	99.2
Si–C	100.5
O–Si–C	101.8
SiO$_x$	103.2
C–Si	283.2
C–C	284.6
C–O–H	286.4
C=O	288.4

Table 3.
Binding energy of different bonds.

The deconvoluted XPS spectrum of SiC/SiO$_2$ (Colloidal) for Si 2p peak and C 1s is shown in **Figure 5**. Similar results were obtained such as, the BE of 103.35, 99.99, and 97.91 eV that were associated with SiO$_x$, Si–C, and Si–Si bonding, respectively, for Si 2p. Fitted curve for C 1s also showed similar results of C–C and C–Si bonding occurring. The O 1s emission was also found at BE of 532.55 eV. The surface chemical composition of Si, O and C obtained from CasaXPS were 20.91, 33.64 and 45.45, respectively.

The deconvoluted XPS spectrum of SiC/SiO$_2$ (Areosil) for Si 2p peak and C 1s is shown in **Figure 7**. The BE was found to be 103.48 and 100.089 eV for Si 2p which are associated with SiO$_x$ bonding and Si–C bonding, respectively. Whereas, fitted curve for C 1s at 284.76 and 282.88 eV of BE was indicating the C–C and C–Si bonding. The O 1s emission was found at BE of 532.55 eV, and the surface chemical composition of Si, O, and C obtained from CasaXPS were 23.59, 34.51, and 41.35%, respectively. There was evident of small amounts of Na 1s present that accounted for the remaining 0.56% of the chemical composition. This was believed to be the result of surface or apparatus contamination.

Figure 8 shows the deconvoluted XPS spectrum of SiC/SiO$_2$ (needle shape) for Si 2p peak and C 1s. The BE was found to be associated with Si–C bonding at 100.61 eV, Si–Si at 98.35 and 103.74 eV related with Si–O bonding, respectively, for Si 2p. Fitted curve for C 1s was also obtained at 284.46 and 283.02 eV of BE was indicating the C–C and the Si–C bonding. In addition to these, the O 1s emission was also found at BE of 532.55 eV. The surface chemical composition of Si, O, and C obtained from CasaXPS were 25.66, 30.97, and 43.37%, respectively.

Figure 9 depicts the deconvoluted XPS spectrum of SiC/SiO_2 (80 nm) for Si 2p peak and C 1s. The BE was found to be associated with Si–C bonding at 100.41 and 103.84 eV related with Si–O bonding, respectively, for Si 2p. On the other hand, fitted curve for C 1s at 284.76 and 282.28 eV of BE was indicating the C–C and the Si–C bonding. In addition to these, the O 1s emission was found at BE of 532.55 eV. The surface chemical composition of Si, O, and C obtained from CasaXPS was 19.82, 37.28 and 42.90 respectively.

The observed shift in the Si 2p peak between the oxide and the carbide is consistent with the expected change in oxidation state. Additional information on the interface was obtained from the C 1s photoelectron spectrum of the nominally bare SiC, far from the interface as shown in **Figure 10**. The narrow second highest energy peak is the expected C 1s signal from bulk SiC [41]. From the observed chemical shifts, we can determine the ionic state of a particular species. Although the incident X-rays penetrate deep into the sample, only electrons emitted from a thin surface layer are detected. The electron escape depth for Si, C, and O ranges from 2.0 to 2.5 nm [42].

3.4 Differential scanning calorimetry (DSC)

Differential scanning calolimetry (DSC) is a thermoanalytical technique by which the energy absorbed or emitted by the material as a function of temperature or time is measured. In present case, we have studied the extruded-polymer nanocomposites to determine the thermal transitions such as the melting (T_m) of a crystalline polymer as shown in **Figure 11**. The DSC data are presented in **Table 4**.

The curve shapes of the neat material and nanocomposites are very similar, with only slight differences in the leading edge of the melt peak, indicating that the neat structure has little effect on the overall phase structure of the material. Above the melting temperature of about 160°C, the polymer is subject to a thermal degradation that is both dependent on time and on temperature. Adding nanoparticles allow the start of the degradation process to shift toward higher temperatures. The increase of the melting temperature of neat PP to PP + SiC was 161.43–166.12°C. Correlation has also been drawn between orientation (crystal perfection) and the DSC melt peak temperature [43].

The crystallization peak temperatures are shown below in **Figure 11**; they follow the same order as the melt peak temperatures, with the neat pp material being more

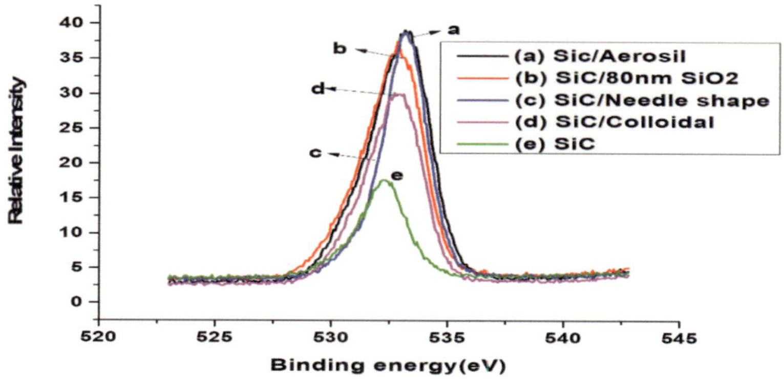

Figure 10.
XPS overlay of O 1s binding energy for SiC and the coated nanoparticles.

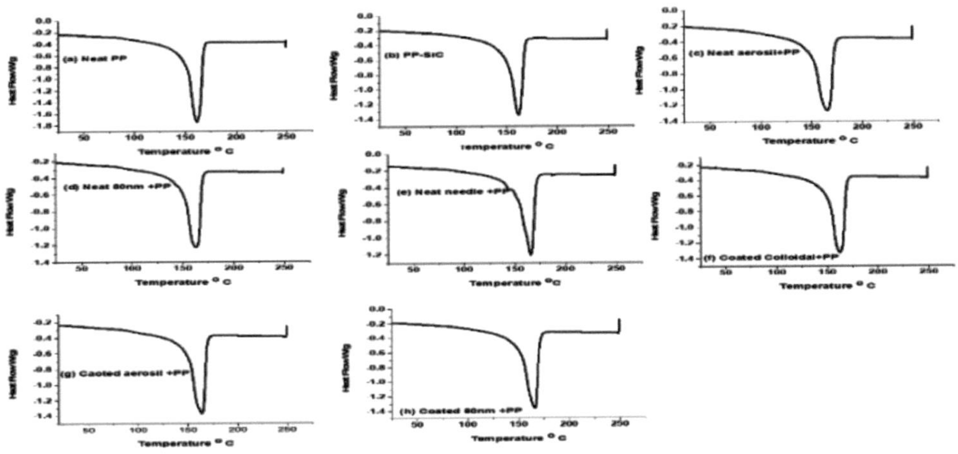

Figure 11.
DSC melting curve of neat PP and nanocomposites.

easily crystallizable. Preliminary isothermal crystallization studies at 116°C of neat polypropylene also show a similar trend. A decrease in the crystallization temperature after introduction of the filler indicates that a higher undercooling of the melt is necessary for the crystallization to occur. Such behavior may be the result of the strong influence of the SiC/SiO_2 nanoparticles on the chain dynamics (lower chain mobility). It has been shown that the crystallization rate increases as both molecular weight and molecular weight distribution increase [44].

Several authors have seen multiple melting peaks in polypropylene, and there is the suggestion of multiple crystalline forms in the polypropylene [33, 44, 45]. It is alternatively suggested that the low temperature melting is caused by a less perfect crystalline order and not from different crystal phases [46]. With very sharp crystallization peaks, it is likely that the low temperature phase is induced at the crystallization temperature. Similar phenomena have been observed for isothermal crystallization [47] and this probably explains the difference in peak temperature between the nanocomposites. It is also plausible that the different polymer fractions do not co-crystallize simultaneously, and that the lower temperature melts is due to morphological effects. It is possible that the higher melt temperature of neat PP implies the molecular weight of the nanocomposites is greater than the neat PP (**Table 4**).

Sample	Melting temperatures (°C)	Crystallization temperatures (°C)
Pure PP	161.43	113.72
Neat areosil	165.48	111.12
Neat 80 nm	163.16	112.83
Neat needle shape	165.60	111.31
PP + SiC	161.51	116.80
Coated areosil	166.12	113.20
Coated colloidal	162.84	113.54
Coated 80 nm	163.70	111.72

Table 4.
DSC melting and crystallization temperature chart.

3.5 Thermogravimetric analysis (TGA)

The applications of TGA are to determine the absorbed moisture content, residual solvent levels, the amount of inorganic filler in a polymer, and in this case the degradation characteristics of the coated nanocomposites in terms of their induction time. The weight loss of a polymer as a function of temperature is commonly determined by this technique. Weight loss of a polymer due to thermal degradation is an irreversible process. Such thermal degradation is largely related to oxidation whereby the molecular bonds of a polymer are attacked by oxygen molecules. **Figure 12** is the TGA graph of extruded neat silicon carbide coating with polypropylene and extruded samples of coated nanoparticles with polypropylene. The TGA data are presented in **Table 5**.

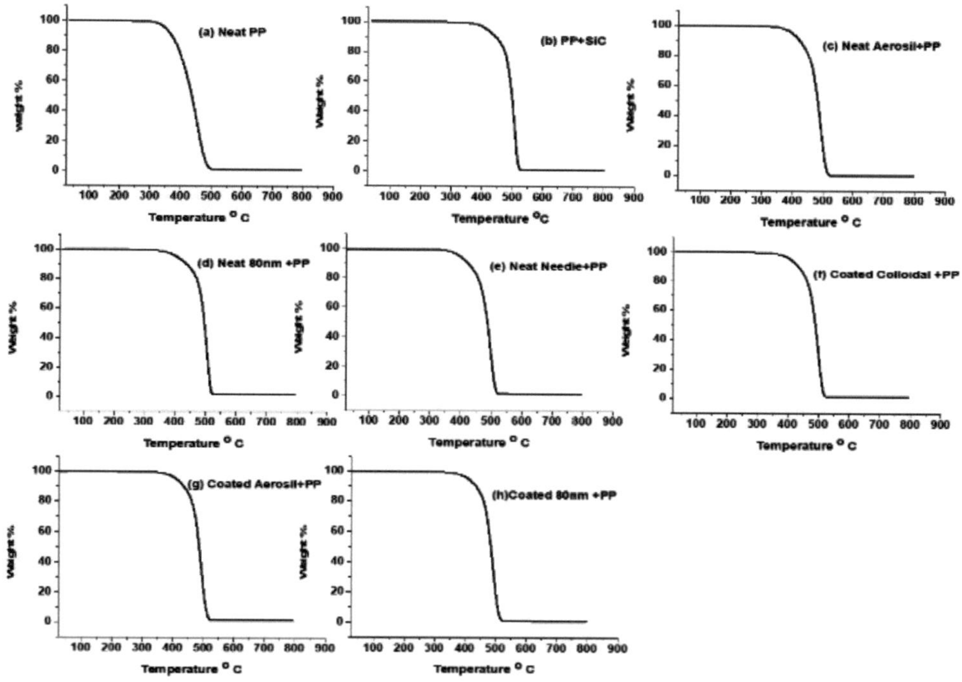

Figure 12.
TGA curves of neat and SiC/SiO$_2$ reinforced polypropylene nanocomposite.

Sample	T_d (°C)	T_{50} (°C)	% Temperature increase
Pure PP	369.78	436.06	—
Neat areosil	452.94	488.59	12.0
Neat 80 nm	469.68	496.34	13.8
Neat needle shape	464.94	493.48	13.1
PP + SiC	453.67	483.60	10.9
Coated areosil	454.06	486.11	11.4
Coated colloidal	446.79	487.50	11.7
Coated 80 nm	447.71	482.39	10.6

Table 5.
Parameters obtained from TGA curves.

In the present study, TGA results show that 50% of the total weight loss is considered as the structural destabilization point of the system. It is a common practice to consider 50% weight loss as an indicator for structural destabilization. From **Figure 12**, it is evident that the neat sample is stable up to 436.06 °C, whereas the coated nanocomposites minimum stabilization temperature is at 482.39°C which is a 10.6% increase. The maximum temperature increase was 13.8% and stabilization occurred at 496.34°C for the neat 80 nm sample. The reason for this increase in the thermal stability is due to the increase in cross-linking of the coated nanoparticles in the presence of polypropylene and having minimum particle to particle interaction.

This may be due to the presence of silicon carbide nanoparticles which are high heat absorbing materials compared to neat polypropylene, which can be the reason for having such a positive impact on improving the decomposition temperature. One can also assume that the nanoparticles were well dispersed and were not acting as impurities in agglomerated form due to the increase. Another probable reason for the observed behavior could be the lower mobility of the polymer chains due to strong interaction with the nanofiller. Thermal degradation of polypropylene begins with formation of the free radicals, which further induce chain fragmentation processes [48]. Volatile oligomers that are formed which are responsible for the mass loss will be partially prevented which results in the whole degradation process to shift towards a higher temperature.

3.6 Flexural test

The flexure test method measures behavior of materials subjected to simple beam loading. It produces tensile stress in the convex side of the specimen and compression stress in the concave side which creates an area of shear stress along the midline.

Flexure stress-strain curves for neat polypropylene and the nanocomposites are shown in **Figure 13**. The flexural test data are presented in **Table 6**. From the graph, it is evident that flexural strength and modulus slightly increases with the addition of coated SiC/SiO_2 nanofillers with respect to the neat polymer. In the table, we see that coated colloidal contributed the maximum percent change in stiffness at 40.5%, while the coated 80 nm contributed to a maximum strength percent change

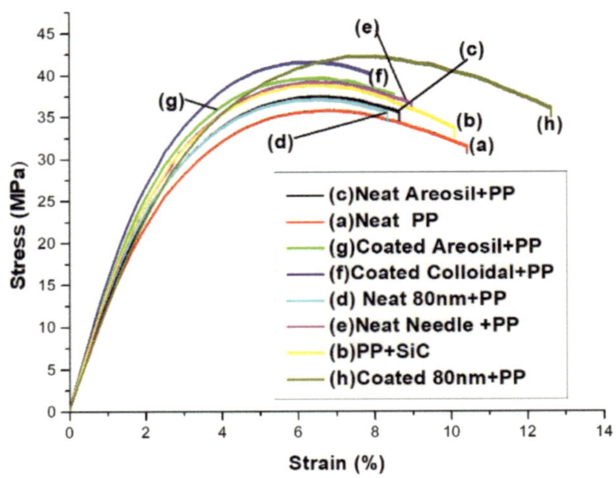

Figure 13.
Stress-strain curve from flexure results.

Sample	Strength (MPa)	% Strength change	Modulus	% Modulus change	Strain (%)	% Increase in toughness
Pure PP	33.97	—	1.16	—	7.09	—
Neat areosil	35.82	5.4	1.31	12.9	6.83	13.2
Neat 80 nm	37.50	10.3	1.42	22.4	6.65	22.4
Neat needle shape	39.76	17.0	1.48	27.5	6.61	27.5
PP + SiC	38.88	14.4	1.45	25	6.50	25
Coated areosil	39.26	15.5	1.49	28.4	6.29	30.1
Coated colloidal	41.65	22.6	1.63	40.5	6.48	41.5
Coated 80 nm	42.33	24.6	1.28	10.3	7.29	92.4

Table 6.
Mechanical properties from flexure test.

at 24.6%, and the greatest toughness percent change which was at 92.4%. Although these were the highest values observed, it is still evident that overall, all of the samples performed greater than the neat PP sample.

The reason for such change in stiffness of the matrix is due to ultrasound irradiation which enhances the homogeneity. The ultrasound irradiation helps in molecular mixing of these components together and the formation of coated reactive particles which ultimately leads to increase the cross-linking in the polymer when mixed. The decrease in mobility of the cross linking in polymer gives better mechanical properties with an optimum amount of nanofiller addition. Surface area also played an important role in the significant increase in that the spherical spatial distribution and low weight ratio of the nanofillers enhance the homogeneity of the reaction. The nanocomposites exhibit ductile fracture and the elongation at break increased with the increase of the nanoparticle content (**Table 6**).

4. Conclusion

The samples were prepared using high intensity ultrasound irradiation to coat SiC onto SiO_2 nanoparticles. This process has been introduced as an efficient way to disperse SiC onto SiO_2 nanoparticles, and by extruding these samples with polypropylene, it increases its mechanical properties. XRD analysis confirmed that the samples were crystalline with an amorphous background present. The TEM analysis revealed that the SiC nanoparticles were uniformly dispersed over the entire surface of the SiO_2 nanoparticles. It also indicated a minor growth in the size of the nanoparticles after ultrasonic irradiation. XPS spectrographs revealed the chemical composites found in the samples. It also verified the surface coating through the shifts in binding energies. The crystallization of the polypropylene polymer matrix was also affected by the presence of the coated nanoparticles. DSC testing indicated that the crystallization and the melting temperatures increased with increasing filler content. Through DSC, it can be seen that by adding nanoparticles, it allows the start of the degradation process to shift toward higher temperatures. The increase of the melting temperature of neat PP to PP + SiC was 161.43–166.12°C. A decrease in the crystallization temperature after introduction of the filler indicates that a higher undercooling of the melt is necessary for the crystallization to occur.

The TGA data showed that the neat sample is stable up to 436.06°C, whereas the coated nanocomposites minimum stabilization temperature is at 482.39°C which is a 10.6% increase. The maximum temperature increase was 13.8% and stabilization occurred at 496.34°C for the neat 80 nm sample. The coated nanocomposites had profound effects on the PP properties. The flexure test results showed that the flexural strength and modulus slightly increases with the addition of coated SiC/ SiO_2 nanofillers with respect to the neat polymer.

Acknowledgements

The authors acknowledge the financial support of NSF-RISE # 1459007, NSF-CREST#1735971 and NSF-MRI-1531934 grants.

Conflict of interest

The authors declare no conflict of interest.

Author details

Vijaya Rangari* and James Davis
Department of Materials Science and Engineering, Tuskegee University, Tuskegee, Alabama, USA

*Address all correspondence to: vrangari@tuskegee.edu

References

[1] Wei S, Xu L, Shen Y, Zhang L, Wu X. Study on microscopic mechanism of nano-silicon dioxide for improving mechanical properties of polypropylene. Molecular Simulation. 2020;**46**(6):468-475

[2] Awad AH, El-Gamasy R, Abd El-Wahab AA, Abdellatif MH. Mechanical behavior of PP reinforced with marble dust. Construction and Building Materials. 2019;**228**:116766

[3] Massoud GB, Carl D, Denis R. Mechanical, water absorption, and aging properties of polypropylene/flax/glass fiber hybrid composites. Journal of Composite Materials. 2015;**49**:3781-3798

[4] Lule ZC, Kim J. Thermally conductive polybutylene succinate composite filled with Si-O-N-C functionalized silicon carbide fabricated via low-speed melt extrusion. European Polymer Journal. 2020;**134**(5):109849

[5] Biswas MC, Tiimob BJ, Abdela W, Jeelani S, Rangari VK. Nano silica-carbon-silver ternary hybrid induced antimicrobial composite films for food packaging application. Food Packaging and Shelf Life. 2019;**19**: 104-113

[6] Akpan EI, Shen X, Wetzel B, Friedrich K. Design and synthesis of polymer nanocomposites. In: Polymer Composites with Functionalized Nanoparticles Synthesis, Properties, and Applications, Micro and Nano Technologies. Elsevier; 2019. pp. 47-83

[7] Damian Beasock T, Stokes M, El-Ghannam A, Schmitz T. Effect of processing parameters on the microstructure and mechanical behavior of a silicon carbide-silica composite. Procedia Manufacturing. 2019;**34**:747-753

[8] Imam MA, Jeelani S, Rangari VK. Thermal decomposition and mechanical characterization of poly (lactic acid) and potato starch blend reinforced with biowaste SiO_2. Journal of Composite Materials. 2019;**53**:2315-2334

[9] Mekuria TD, Zhang C, Fouad DE. The effect of thermally developed SiC@ SiO_2 core-shell structured nanoparticles on the mechanical, thermal and UV-shielding properties of polyimide composites. Composites Part B: Engineering. 2019;**173**:106917

[10] Abdel-Hamid IM. Thermo-mechanical characteristics of thermally aged polyethylene/polypropylene blends. Materials and Design. 2010;**31**:918-929

[11] Greco R, Ragosta G. Isotactic polypropylene of different molecular characteristics: Influence of crystallization conditions and annealing on fracture behavior. Journal of Materials Science. 1988;**23**:4171-4180

[12] Ouderni M, Philips PJ. Influence of morphology on the fracture toughness of isotactic polypropylene. Journal of Polymer Science. Part B. 1995;**33**:1313-1322

[13] Thumm A, Risani R, Dickson A, Sorieul M. Ligno-cellulosic fibre sized with nucleating agents promoting transcrystallinity in isotactic polypropylene composites. Materials. 2020;**13**:1259

[14] Ščetar M, Kurek M, Režek JA. Effect of high power ultrasound on physical–chemical properties of polypropylene films aimed for food packaging: Structure and surface features. Polymer Bulletin. 2019;**76**:1007-1021

[15] Liang J-Z. Effects of tension rates and filler size on tensile properties of polypropylene/graphene nano-platelets

composites. Composites Part B Engineering. 2019;**167**:241-249

[16] Kulkarni MB, Radhakrishnan S, Samarth N, Mahanwar PA. Structure, mechanical and thermal properties of polypropylene based hybrid composites with banana fiber and fly ash. Materials Research Express. 2019;**6**:075318

[17] Ibhadon O. Fracture mechanics of polypropylene: Effect of molecular characteristics, crystallization conditions, and annealing on morphology and impact performance. Journal of Applied Polymer Science. 1998;**69**:2657-2661

[18] Friedrich K. Strength of crystalline isotactic polypropylene and the effect of molecular and morphological parameters. Progress in Colloid and Polymer Science. 1979;**66**:299-309

[19] Van der Wal A, Mulder JJ, Gaymans RJ. Fracture of polypropylene: The effect of crystallinity. Polymer. 1998;**39**:5477-5481

[20] Wetzel B, Haupert F, Zhang MQ. Epoxy nanocomposites with high mechanical and tribological performance. Composites Science and Technology. 2003;**63**:1-13

[21] Zheng Y, Zheng Y, Ning R. Effect of nanoparticles SiO_2 on the performance of nanocomposites. Materials Letters. 2003;**57**:2940-2944

[22] Ying W, Jing Z, Xinyuan S, Changnian S, Jiajun W, Li S. Dispersion investigation of TiO_2 nanoparticles coated by pulsed RF plasma polymer. Materials Chemistry and Physics. 2006;**98**:217-224

[23] Dai H, Wong EW, Lu YZ, Fan SS, Lieber CM. Synthesis and characterization of carbide nanorods. Nature. 1995;**374**:769-772

[24] Bralla JG, editor. Handbook of Product Design for Manufacturing. New York: McGraw-Hill Publishing Company; 1986

[25] Wick C, Benedict JT, Veilleux RF. Tool and Manufacturing Engineers Handbook. 4th ed. Vol. 2. Society of Manufacturing Engineers (SME); 1984. ISBN: 0-87263-135-4

[26] Hassan M, Zhou Y, Mahfuz H, Jeelani S. Effect of SiO_2 nanoparticles on thermal and tensile behavior of nylon-6. Materials Science and Engineering A. 2006;**429**:181-188

[27] Sung-Po L, Shyh-Shin H, Jui-Ming Y, Chi-Chang H. Mechanical properties of polyamide-6/montmorillonite nanocomposites. Prepared by the twin screw extruder mixed technique. International Communications in Heat and Mass Transfer. 2011;**38**(1):37-43

[28] Russo GM, Nicolais V, Di Maio L, Montesano S, Incarnato L. Rheological and mechanical properties of nylon 6 nanocomposites submitted to reprocessing with single and twin screw extruders. Polymer Degradation and Stability. 2007;**92**:1925-1933

[29] Yong L, Haibin C, Chi-Ming C, Jingshen W. Effects of coating amount and particle concentration on the impact toughness of polypropylene/$CaCO_3$ nanocomposites. European Polymer Journal. 2011;**47**:294-304

[30] Manolis GD, Gareis PJ, Tsono AD, Neal JA. Dynamic properties of polypropylene fiber-reinforced concrete slabs. Cement and Concrete Composites. 1997;**19**(4):341-349

[31] Xiangmeng L, Ming K, Lu Y, Simin S, Rong S, Lixian S, et al. Quantitative evaluation of fillers dispersion state in $CaCO_3$/polypropylene composites through visualization and

fractal analysis. Polymer Composites. 2020;**41**:1605-1613

[32] Vijaya KR, Tarig AH, Quentin M, Shaik J. Size reduction of WO_3 nanoparticles by ultrasound irradiation and its applications in structural nanocomposites. Composites Science and Technology. 2009;**69**:2293-2300

[33] Vijaya KR, Mohammad SB, Shaik J. Microwave processing and charactering of EPON 862/CNTs nanocomposites. Materials Science and Engineering B. 2010;**168**:117-121

[34] Margiotta JC. Study of silicon carbide formation by liquid silicon infiltration of porous carbon structures [PhD dissertation]. Baltimore, MD, USA: The Johns Hopkins University; 2009

[35] Solomon I, Schmidt MP, Senemand C, DrissKhodja M. Band structure of carbonated amorphous silicon studied by optical, photoelectron, and X-ray spectroscopy. Physical Review B. 1988;**38**:13263

[36] Lee WY. X-ray photoelectron spectroscopy and Auger electron spectroscopy studies of glow discharge $Si_{1-x}C_x$:H films. Journal of Applied Physics. 1980;**51**:3365

[37] Choi WK, Loo FL, Loh FC, Tan KL. Structural and electrical studies of radio frequency sputtered hydrogenated amorphous silicon carbide films. Journal of Applied Physics. 1995;**78**:7289

[38] Demichelis F, Giorgis F, Pirri CF, Tresso E. Bonding structure and defects in wide band gap a-$Si_{1-x}C_{x:H}$ films deposited in Hz diluted SiH_4 + CH_4 gas mixtures. Philosophical Magazine B. 1995;**71**:1015

[39] Demichelis F, Crovin G, Pirri CF, Tresso E, Rava P, Gallini R, et al. Optimization of relevant deposition parameters for high quality a-SiC:H films. Solar Energy Materials & Solar Cells. 1995;**37**:315

[40] Katiyar M, Yang YH, Abelson JR. Si–C–H bonding in amorphous $Si_{1-x}C_x$:H film/substrate interfaces determined by real time infrared absorption during reactive magnetron sputter deposition. Journal of Applied Physics. 1995;**78**:1659

[41] Muehlhoff L, Choyke WJ, Bozack MJ, Yates JT. Comparative electron spectroscopic studies of surface segregation on SiC(0001) and SiC(0001). Journal of Applied Physics. 1986;**60**:2842

[42] Himpsel FJ, McFeely FR, Taleb-Ibrahimi A, Yarmoff JA. Microscopic structure of the SiO_2/Si interface. Physical Review B. 1988;**38**:6084

[43] Jaffe M, Turi E, editors. Thermal Characterization of Polymeric Materials. New York: Academic Press; 1981. pp. 731-781

[44] Ahmed M. Polypropylene Fibers - Science and Technology. New York: Elsevier; 1982. p. 187

[45] Saraf AW, Desai P, Abhiraman AS. Applied Polymer Symposium 47. New York: Wiley; 1991. p. 67

[46] Caldas V, Brown GR, Nohr RS, Macdonald JG, Taboin LE. The structure of the mesomorphic phase of quenched isotactic polypropylene. Polymer. 1994;**35**:899

[47] Crompton TR. Thermal Methods of Polymer Analysis. Smithers Rapra Technology; 2013. ISBN-10:1847356621

[48] Allen NS, Edge M. Fundamentals of Polymer Degradation and Stabilization. 2nd ed. London: Elsevier Applied Science; 1992

Chapter 3

Advanced Polypropylene and Composites with Polypropylene with Applications in Modern Medicine

Doina Elena Gavrila, Victor Stoian, Alina Caramitu and Sorina Mitrea

Abstract

Synthetic polypropylene (PP) is used extensively in many fields of medicine. On the one hand, it is utilized in the manufacture of medical equipment: syringes, storage, transport, electric cables, etc. On the other hand, synthetic, nonabsorbable isotactic PP (iPP) is often used to perform meshes for hernia and pelvic organ repair operations, as well as in urinary incontinence. Products that release in time from meshes are depending on the conditions in which they are utilized, can produce undesirable reactions for the human body. For this reason, nonabsorbable synthetic PP was replaced in surgical sutures and meshes with bio polypropylene (bio PP). The chapter analyzes the specific characteristics of these polymers as well as their degradation due to the influence of different factors: humidity, perspiration, temperature, and presence of bacteria. Obtaining new composite materials with PP as matrix and metal powders as fillers is considered as a possibility of their use in vital problems such as cancer detection and treatment. These allow the emergence of new strategies in the design of biosensors that use nanocomposite materials with different fillers and polymeric films. The chapter analyzes the characteristics of new composite materials with PP matrix and metallic powders of iron (Fe).

Keywords: polypropylene, bio polypropylene, degradation, surgical sutures, polypropylene composite with iron powders

1. Introduction

Polypropylene (PP) is a nonpolar, partially crystalline thermoplastic polymer from the polyolefin group, in some aspects similar to polyethylene (PE) concerning the electrical properties, but it has improved mechanical properties and thermal resistance. In these latter days, PP is the most used thermoplastic, exceeding PE [1]. The properties of PP depend on molecular weight and molecular weight distribution, crystallinity, and tacticity [2]. It is a low-cost polymer with excellent properties like flame resistance, transparency, high heat distortion temperature, dimensional stability, and recyclability, making it ideal for a wide range of applications. It is a lightweight polymer with a density of 0.90 g/cm^3 that makes it suitable for many industrial applications and has chemical resistance, long life for electrical and mechanical applications, and thermal and optical properties. It is environment-friendly and nontoxic and has replaced polyvinylchloride (PVC) because it does not release any toxic gases when burnt [3]. PP can be modified by using micro- and nano-sized fillers or reinforcing agents to get desired characteristics required by the application area. The tacticity of PP can be controlled: the branching of linear PP can be done, resulting in higher molecular weight product with higher tensile strength, higher modulus of rigidity, and higher heat resistance. PP does not absorb water [4].

2. Biopolymers and biopolypropylene

Biopolymers are polymers produced by living organism and contain monomeric units that are covalently bonded to form larger structures. A classification according to monomeric units and the structure of the biopolymer divide these materials into three classes:

- Polynucleotides (DNA and RNA) which are long polymers composed of 13 or more monomeric units covalently bonded in a chain. DNA (deoxyribonucleic acid) and RNA (ribonucleic acid) are examples of polynucleotides with distinct biological function. DNA consists of two chains of polynucleotides, each in the form of a helical spiral.

- Polypeptides, which are short polymers of amino acids.

- Polysaccharides, which are often linear-bonded polymeric carbohydrate structures.

Other examples of biopolymers are rubber, melanin, lignin, cellulose, etc.

An important difference between polymers and biopolymers is their structure. All polymers are made of repetitive units called monomers. Although biopolymers have a well-defined structure, this is not a defining characteristic [5–7].

Structural biology is the domain for the study of structural properties of biopolymers. The exact chemical composition and the sequence in which the units are arranged are called the primary structure, in the case of proteins. Many biopolymers fold into a characteristic compact shape, as well as secondary and tertiary structure. These structures determine the biological functions and depend in a complicated way on their primary structures. Most synthetic polymers have much simpler and more random structures. Molecular mass distribution is missing in biopolymers: all types of biopolymers contain similar sequences and numbers of monomeric units. The phenomenon is called monodispersity, in contrast to the polydispersity from synthetic polymers. Biopolymers have a polydispersity index of 1.

Synthetic biopolymers are human-made copies of biopolymers obtained by abiotic chemical routes. Abiotic components or abiotic factors are nonliving chemical and physical parts of the environment that affect living organism and the functioning of ecosystems. In biology, abiotic factors can include water, light, radiation, temperature, humidity, atmosphere, acidity, and soil. The macroscopic climate often influences each of the above [8, 9].

As humanity becomes more aware of climate change and pollution with plastics, the idea of using biopolymers is becoming more and more important. At this time, alternative methods of producing biopolymer materials are however

-Conventional petrochemicals method →Crude oil →Naphta→ →Propylene polymerization →PP

-Bio method →Raw biomass materials →bio Ethanol →bio Propylene →bio PP

-Bio method →Raw biomass materials (fermentation) →bio Isopropanol (dehydratation) →bio Propylene →bio PP

Figure 1.
Scheme of methods for obtaining PP and bio PP.

reduced. It is necessary to accelerate the transition from petrochemical to bio-based plastic materials and to obtain a great diversity of them, including polyolefins as PP and PE [10]. Recently, in the bio-based polymer industry, the world's first parallel production of bio-based PP and bio-based low-density PE at the commercial scale has been announced. However, due to the technical difficulties, production of bio PP from biomass has not been established at the industrial level. The new production method is proposed which sees various biomasses, mainly nonedible plants fermented to produce isopropanol (IPA), which is then dehydrated to obtain propylene. This one could prove to be a more cost-effective way to manufacture bio PP. **Figure 1** shows schematic methods of obtaining PP and bio PP.

Biopolymers remain less than one percent of plastics manufactured worldwide and remain until very recently 2–4 times more expensive than petrochemical products. Moreover, most biopolymers do not yet save more carbon emissions than are required to manufacture them. As most biopolymers are produced from plants, sugars, starches, or oils, it was estimated that replacing 250 million tones of the plastic manufactured each year with bio plastics would require 100 million hectares of land (7% of the arable land of the Earth).

3. Analysis of degradation of bio polypropylene for surgical sutures and for meshes used in medical repair operations

3.1 Generalities on suture materials

Surgery sutures have improved in recent years, both in terms of surgical technique and materials used, which explains successes in transplant, vascular, and digestive surgeries, etc. Initially, sutures were made from natural biological materials, modern materials being synthetic, mostly polymers. Suture materials are absorbable (resorbable) or nonabsorbable (non-resorbable) and monofilamentous or multifilamentous (braided sutures), their diameter being a very important factor. Generally, fast-healing tissues and internal organs are treated with resorbable materials, while slow-healing tissues and tissues with high mechanical exposure, such as the skin and tendons, are treated with non-resorbable materials. Synthetic suture materials are absorbed by hydrolysis due to water whose molecules attack the ester linkages and break up the polymer chains. The degradation for biological suture materials as the same for synthetic suture materials begins with the breaking down of polymer chains into smaller fragments. The fragments are then phagocytized by the enzymatic action of special types of mononuclear and multinuclear white blood cells [11, 12]. The idea of using sutures as a medicament transport system is a topic of interest for modern surgery, reducing the risk of infection and inflammation. Bio PP is mainly absorbed by hydrolysis due to water, the amount of water being essential for the rate of hydrolysis reaction.

3.2 Materials

Bio PP from BioSintex is a monofilament sterile suture made of isotactic PP (iPP). The studied bio PP materials were blue and had the thickness of.

1 EP (European Pharmacopoeia) and 3.5 EP, respectively. The degradation of bio PP mesh designed for surgical treatment of urinary incontinence (white color, with monofilament diameter 0.10 mm, typical size pore 1.7 × 0.3 mm, thickness 0.50 mm) was studied.

3.3 DSC analyses.

Measurements have been made with the Calorimeter 200F3, Maia, Netzsch-Geratebau GmbH with a measuring cell: gas flow control device (N_2); sealing press; heating rate 10°C/min, sample mass approximately 10 mg, sample pan aluminum. Temperature range was −30 to 200°C. One scan was used in DSC analysis. The crystallinity determination was performed using analysis software Netzsch TA—Proteus.

3.4 Preparation of solutions: methods

Perspiration consists of water, minerals, lactate, and urea. On average, the mineral composition is sodium (0.9 g/l), potassium (0.2 g/l), calcium (0.015 g/l), and magnesium (0.0013 g/l). We prepared a sodium solution in water with a concentration of 0.9 g/l. A water solution was prepared in which beechwood was introduced. This is because experimental measurements have shown the existence of bacteria in beechwood. In these experiences there are two categories of degradation media: water and salt or beechwood solutions in water. In both cases we made measurements by introducing samples 1, 2, 3, 4, ..., 40 days, ..., 60 days in the two categories of environments. In the water important changes appeared 42 days, while in the solutions with salt, changes appeared after 4 days.

3.5 Results

For bio PP blue, thickness 1 EP, which has not been degraded, the initial crystallinity of the material is 55.91%, the melting point is T_m = 168.7°C, and the corresponding value is 1.247 mW/mg. DSC curve for blue bio PP, thickness 1 EP, kept for 6 days continuously in water shows that there are no fundamental changes in this case: the melting temperature, T_m, remains the same, and there is a small decrease from 1.247 mW/mg to 1.154 mW/mg. There is a very low increase in crystallinity: from 56.40 to 56.87%. What does this little increase in crystallinity show? It is known that the penetration of the water takes place initially in the amorphous area, it destroys by the phenomenon of hydrolysis and then enter in the crystalline area. For bio PP, after 6 days, the water entered the amorphous zone and destroyed a part of it. This shows that suture materials studied retain practically their mechanical characteristics.

Measurements made on white meshes from bio PP and DSC characteristics show changes identical to those that occur in iPP. The crystallinity of the meshes was 52.24%, melting temperature was 170.5°C, and has the value of 1.044 mW/mg. The bio PP meshes originally introduced for the repair of the hernia are currently used in several anatomical sites. From millions of meshes implanted annually around the world, more than 10% are removed due to complications. The fundamental question of degradation of PP in vivo has not been elucidated [13, 14]. We consider that in vivo there are a number of factors such as acid solutions, salted solutions, and solutions with bacteria and enzymes that produce a strong destruction of PP. As will be seen from the analyses, the salt solutions destroy bio PP in a short time, affect the crystalline system, and consequently cause the destruction of the material. Substances that are removed can affect the body and cause complications. According to the data from the literature the degradation products are acids and alcohols from PP degradation due to the action of the water [11].

Figure 2 shows DSC curve for bio PP blue, thickness 3.5 EP, which has not been degraded, and the DSC curve for the same material that was kept in water for 42 days. There is a very slight increase in crystallinity in this case by degradation: from 63.08% to 63.84%. There is a very low melting temperature decrease from 165.8°C to 165.7°C but also a very small increase from 1.051 mW/mg to 1.15 mW/mg. What is most important in this case is the appearance of a second additional maximum melting at 154.8°C, indicating a second important type of crystalline lamellas that appeared due to the material degradation. The lower melting temperature indicates that these lamellas melt more easily.

To study the influence of perspiration on bio PP sutures, samples have been kept in sodium solution in water for 4 days. In this case there are more important changes in the melting temperature and in the decrease of crystallinity (**Figure 3**).

The discontinuities that occur are due to the melting of various crystalline lamellae at different temperatures, indicating a degradation of crystalline domains. As shown in **Figure 3**, there are several Flory components, each of these components having a single active crystallization center. The T_m values as well as crystallinity decrease. The melting point decrease may be related to the existence of chain branches. Below T_m, the amorphous regions alternate with regions which are lamellar crystals. The amorphous regions contribute elasticity, and the crystalline regions contribute strength and rigidity.

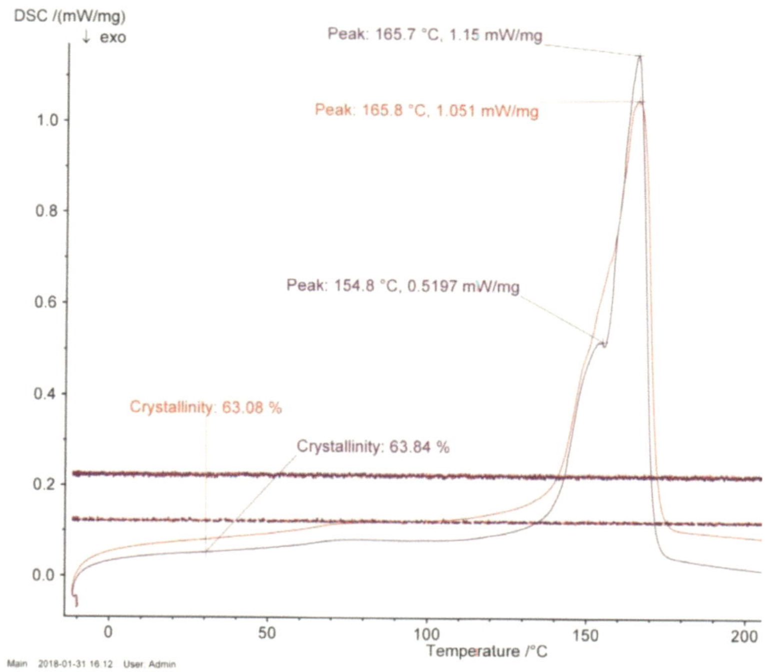

Figure 2.
DSC dependencies for non-degraded blue bio PP, 3.5 EP, (red color) and blue bio PP, 3.5 EP, kept 42 days continuously in water.

For DSC curves of bio PP, 3.5 EP, which was kept for a longer time in water, 60 days, melting point discontinuities and an increase in crystallinity are observed (**Figure 4**).

Outside the main peak at 168.6°C, a very close maximum is observed at 164.7°C, the value 1.325 mW/mg being higher than those obtained for non-degraded material. An increase in crystallinity was obtained: from 63.08% to 66.40%. As was observed in analyzes, bio PP fibers show very good water resistance being considered as high-performance materials in this field. This confers a high thermal resistance keeping the elasticity and mechanical resistance characteristics at

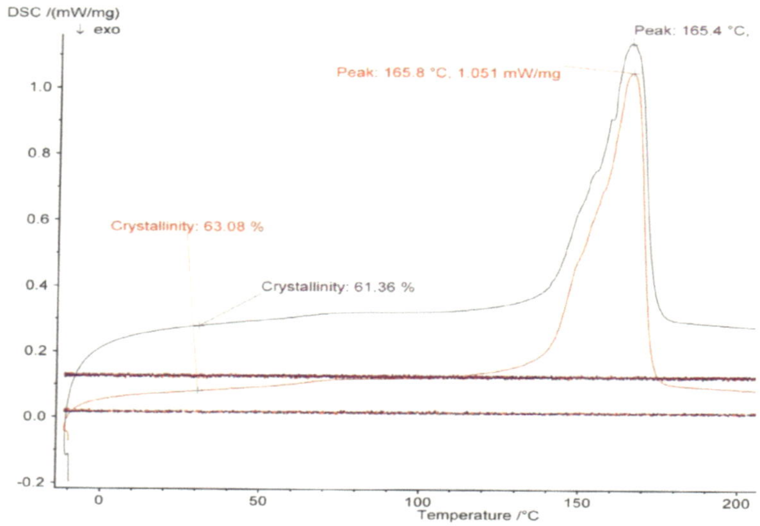

Figure 3.
DSC curve for blue bio PP, thickness 3.5 EP, (red color) and DSC curve for blue bio PP, 3.5 EP, kept continuously 4 days in a sodium solution in water (mauve color).

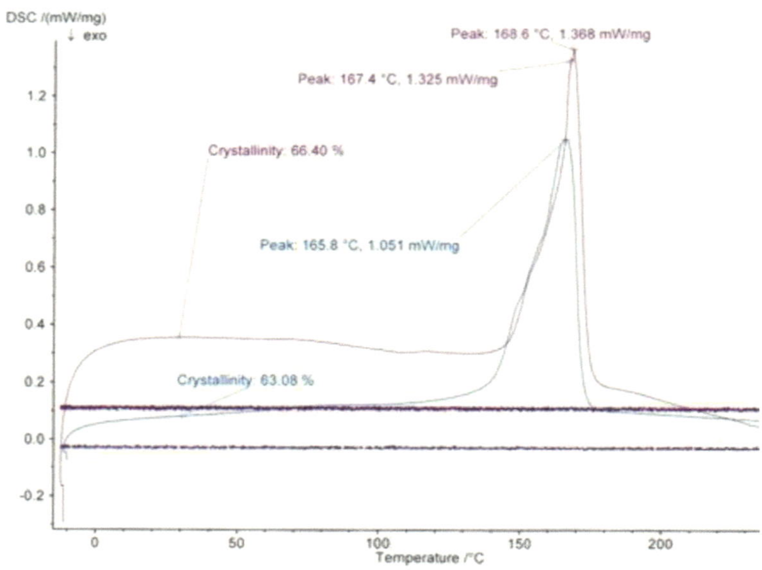

Figure 4.
Comparison between DSC curves for blue bio PP, 3.5 EP undegraded, (blue curve) and the blue bio PP, 3.5 EP, 60 days in water (mauve curve).

temperature up to 90°C and good resistance to acids and alkalis. The polymer is depolymerized by the conversion of the polymer chains into a monomer or mixture of monomer, oligomers, dimers, etc. Regarding the samples that were introduced into the water with beechwood, reduced degradation was obtained compared to those introduced only in water.

At this time, sutures with PP and PE composite filaments allow the transport of drugs for different treatments.

4. Effect of powders of iron on the properties of isotactic polypropylene

4.1 Composites for biosensors

Early diagnosis of cancer is crucial for successful treatment of this disease. Highly sensitive methods are needed for measuring cancer diagnosis markers present at very low levels during early stages of the disease [15]. Major progress in the field of nanotechnology for the creation of small integrated and reliable nano-transducers in combination with biological detection elements has revolutionized the field of biosensors in the last decade. Such biosensor systems allow obtaining biochemical and biophysical signals associated with a particular disease at the level of a single molecule or cell. At present, cancer can be diagnosed mainly by monitoring certain antigens in the blood stream or other fluids of the human body or by histological examination. Modification of electrodes with nano-materials is a major interest for science and nanotechnology because they can be used in different fields in this regard. However, the immobilization strategies of biomolecules are a challenge because their specific properties must be maintained after immobilization in order to obtain reliable sensors [16]. The sensitivity of the sensors can be increased by using nano-materials with polymeric matrix and metallic fillers. Innovative biosensor strategies would allow cancer testing to be performed more rapidly, inexpensively, and reliably in a decentralized setting.

The composite materials with polymeric matrix and metallic fillers, in particular nanoparticles, present the modification of their general characteristics. The development and continuous improvement of the technology for producing and molding the above materials is accompanied by research on their mechanical, electrical, and thermal properties [17–19]. These characteristics as well as changes in moisture absorption, glass transition temperatures, flame retardant, solubility etc. must be given special attention. Interesting and intense studies refer to these materials. It is therefore possible to replace the traditional materials with new composite lightweight materials, and their use can be practicable in many fields.

A novel pan-milling technique was developed to prepare ultrafine PP-Fe composite powders, in which the average grain size of the Fe particles attain a nanoscale level [20]. The authors of the article [21] compose a detailed review to provide a description of the recent advances in the preparation of polymer nanocomposites via mechanical milling.

Mechanical properties of metal-PP polymer matrix composites were investigated experimentally [22]. High-density polyethylene (HDPE), PP, and polystyrene (PS) were used as polymer matrix, and Fe powder in 5, 10, and 15 vol% was used as metal. The modulus of elasticity, yield and strength, elongation, Izod notched impact strength, Shore D hardness, and fracture surfaces of the composites were determined. Addition of 5 vol% reduced Izod impact strength and increased the modulus of elasticity. Thermal stability of polymer nanocomposites prepared by melt mixing of PP with organo-modified clay and maleic anhydride-grafted PP has been studied [23].

Electrical properties of PP-Fe-modified and PP-Fe-nonmodified powders were studied [18]. The phenomenon of percolation and the increase of the electrical conductivity were obtained by two methods of preparation: ball milling and extrusion route. The extrusion resulted in a good dispersion of Fe particles in the PP matrix.

4.2 Materials

The polymer used as matrix was isotactic polypropylene (Tipplen). Conductive reinforcements used were two types of iron powders from Nanografi LTD. Histogram granulometric distribution of iron powder 50 nm showed the existence of a continuous distribution of particles with a maximum around 140 nm and a sharp decline of the number of particles after this value [24]. From SEM analysis for Fe powders, it was found that the mean particle size was 82.2 nm (calculated from eight measurements) for the first powder of 50 nm and the mean size (calculated from 6 measurements) of the particles of the second powder of 800 nm was 2695 nm [24, 25].

4.3 Sample preparation

PP composite samples with iron nano-powders were prepared by extrusion. Samples were obtained in two stages. In the first step, the components (PP and iron powder) were blended for 1 h in a Turbula T2F cylindrical mixer with 1.3 l mixing bowl and with clamping device from rubber, having a rotation speed of 40 rpm. By extrusion (Brabender Ketse laboratory extruder) the composite granules were obtained. In the second step, the composite granules were injected (Dr Boy A35 injection machine); thus the samples disk shaped are obtained with a diameter of 30 mm and a thickness of 2.5 mm. Polymeric composite materials with increasing content of powders (3, 5, and 8%) were prepared.

The working parameters on the extruder were extruder screw speed (45 rpm) and power funnel screw speed (700 rpm). The temperatures on the extruder heating areas in five zones were 170, 175, 180, 190, and 200°C.

4.4 Experimental

4.4.1 Mechanical tests: microindentation hardness and Shore hardness testing

Indentation hardness tests are used in mechanical engineering to determine the hardness of a material to deformation. In microindentation hardness the testing of materials is made with low applied loads, typically of 2N (roughly 200 gf). The Vickers (HV) hardness and Young's (E_{IT}) module were determined by instrument microinjection tests according to ASTM E2546-15 and the Oliver-Pharr calculation method. Mechanical tests were performed at ambient temperature, relative air humidity 35 ± 5%, with Micro-Combi equipment.

4.4.2 Thermal conductivity tests

Thermal diffusion measured between 25°C and 95°C was determined using a "flash" method (ASTM E-1461: 2007), with a LFA 447 NanoFlash (Netzsch). As a radiation source, a xenon performance lamp was used, and the irradiation time on the front face of the sample was 0.18 ms. Samples were analyzed three times at each temperature. The increase in temperature on the other surface of the sample was measured with an InSb-type infrared (IR) detector. The sample analyzed had cylindrical shape Ø = 12.7 mm and height h = 1.95 mm, in pressed stat, and was analyzed at 25°C. Samples were grafted into suspension. Working conditions are as follows: the data acquisition software used was Nanoflash, temperature/°C was 25.0, analysis type was single layer, the analysis software used was Proteus LFA, and the analysis method used was Cowan model.

4.4.3 Electrical conductivity tests

It is known that the thermal conductivity and the electrical conductivity are in a close correlation. The dielectric tests are performed by the Solatron IL MAv-0.6 269 ... + 400°C impedance analyzer; temperature rise rate (heating/cooling) was 0.01—30°C/min; thermal stability was max ±0.01°C; frequency domain was 10 µHz ... 20 MHz; thermal stabilization was max. 8 min; the data acquisition software used was Smart. Tests were conducted in air at 3 V. The samples had a diameter of 30 mm and an average thickness of 4.3 mm. The electrical conductivity in alternating electrical current was determined.

4.4.4 DSC analysis

The conditions of measurements are described in Section 3.3.

4.4.5 SEM analysis

Electron beam scanning microscope with field emission source and ion-focused beam was used. The images were made at 1 or 2 kV acceleration voltage with a very large lens approach. The detector used was that of the Everhart-Thornley secondary electron with the Faraday cup, resulting in micrographs that highlight the morphology and topography of the analyzed surfaces.

4.5 Results and discussions

4.5.1 Analyzing the situation of reinforcement of the polymeric materials

Analyzing the situation of reinforcement of the polymeric materials with different filler materials from **Table 1**, it could be concluded that the use of dispersing fillers would lead to a decrease in the composite resistance compared to the non-filler polymer. This observation is susceptible to interpretations, because **Table 1** shows in all cases an increase in the modulus of elasticity E_{IT}, the energy of the reversible elastic deformation $W_{elastic}$, and the yield of the elastic zone η_{IT} (%), which indicates an improvement in the elastic characteristics. The addition of rigid filler particles produces an increase in the modulus of elasticity which is proportional to the volume percentage of the filler. The worsening of mechanical properties can be attributed to the occurrence and development of cracks in the vicinity of the filler particles and the formation of agglomerates of filler material. These agglomerates are initiated by the voids resulting from the partial detachment of the matrix.

Shore durometer scale D was used to measure the hardness and to characterize the penetration resistance of a penetrator in the composite materials studied. The obtained results are presented in **Table 2**.

All the materials studied have a higher Shore hardness than pure iPP. It is noted that the higher Shore hardness values are obtained for the composite iPP-Fe 82.8 nm.

	H_{IT} (MPa)	HV	E_{IT} (GPa)	h_{max} (ηm)	$W_{elastic}$ (μJ)	W_{total} (μJ)	η_{IT} (%)
iPP	114.7 ± 17.1	10.8 ± 1.6	1.4 ± 0.16	23.7 ± 0.5	2.4 ± 0.9	8.3 ± 0.1	28.6 ± 1.1
iPP with 8% Fe 2695 nm	104.3 ± 14.1	9.8 ± 1.3	1.4 ± 0.07	23.9 ± 0.9	3.2 ± 0.1	8.7 ± 0.1	37.0 ± 0.1
iPP with 3% Fe 82.2 nm	110.7 ± 5.6	10.4 ± 0.5	1.4 ± 0.01	23.5 ± 0.1	3.1 ± 0.4	8.8 ± 0.01	34.9 ± 5.2
iPP with 8% Fe 82.2 nm	115.8 ± 9.6	10.9 ± 0.9	1.5 ± 0.07	22.9 ± 0.2	3.0 ± 0.4	8.8 ± 0.1	34.4 ± 4.9

Table 1.
HIT is indented hardness HV is Vickers hardness E_{IT} is elastic modulus h_{max} is the maximum depth of penetration $W_{elastic}$ and W_{total} are the energies of the reversible elastic deformation and the total deformation, respectively and η_{IT} (%) is the yield of the elastic zone.

iPP	Powders (%)	iPP with Fe 82.8 nm	iPP with Fe 2695 nm
55.1	3	67.2	56.1
	5	70.0	62.5
	8	72.8	65.2

Table 2.
Shore hardness scale D for the composite materials studied.

λ, W/(mK) iPP	Powders (%)	λ, W/(mK) iPP with Fe 82.8 nm	λ, W/(mK) iPP with Fe 2695 nm
0.127	3	0.127	0.14
	5	0.129	0.145
	8	0.134	0.14

Table 3.
Dependence of the thermal conductivity (λ) on the concentration of metallic powders.

4.5.2 Thermal conductivity

Thermal conductivity, λ, characterizes the ability of a material to transmit heat when subjected to a temperature difference (**Table 3**). The thermal resistance is mainly caused by phonon scattering processes, including phonon-phonon scattering, boundary scattering, and defect or impurity scattering [26, 27]. In composite materials, phonon scattering is mainly due to the existence of an interfacial thermal barrier, which results from the acoustic mismatch or the damage of surface layer between the filler and polymer matrix. In order to lower the thermal resistance or enhance the thermal conductivity of the composite, measures should be taken to reduce the interfacial thermal barrier, which is closely related to the filler dispersion and the filler-matrix interaction. Many theoretical studies have studied the thermal conductivity dependence of composite polymers on the amount of metal filler or on the volume fraction of the filler, emphasizing the increase in thermal conductivity with the increase in the volume of the filler [28]. Experimental studies in the inorganic filler/polymer composites have shown that the thermal conductivity increases with the filler content, but very high filling load is used to obtain high thermal conductivity. Large amounts of filler affect and worsen the mechanical properties. Filler size and shape are important factors that influence thermal conductivity and other properties. At the same particle size, smaller particle size leads to a lower interparticle distance and more chances for the formation of the thermal conductive pathway. A significant increase in the thermal conductivity of iPP filled with combined filler, with total constant filler content of 7.5%, was highlighted [29].

The thermal conductivity values of the composites are very close. The lowest thermal conductivity is that of pure iPP and for iPP with 3% concentration of powders with size of 82.8 nm. But other observations can be made: it is observed that the largest increase in values of thermal conductivity is obtained for 8% concentration of Fe powders with size of 82.8 nm. It is very important that this composite has the best thermal endurance [25]. The thermal conductivity values show approximately the same increase for 3, 5, and 8% concentration for Fe powders with size of 2695 nm.

4.5.3 The electrical conductivity

The electrical conductivity in alternating electrical current was determined (**Table 4**). It is known that the thermal conductivity and the electrical conductivity are in a close correlation.

iPP	Powder (%)	iPP and Fe 82.2 nm	iPP and Fe 2695 nm
σ (Ω^{-1} m^{-1})		σ (Ω^{-1} m^{-1})	σ (Ω^{-1} m^{-1})
1.2×10^{-4}	3	1.49×10^{-3}	1.46×10^{-3}
	5	1.53×10^{-3}	1.49×10^{-3}
	8	1.56×10^{-3}	1.53×10^{-3}

Table 4.
Dependence of the electrical conductivity σ (Ω^{-1} m^{-1}) of iPP composites with 3, 5, and 8% concentration of Fe powders for ~2 × 10^4 Hz.

Concentration (%)	Powder size (nm)	Crystallinity (%)	Tm (°C)
iPP	—	53.46	170.4
3	Fe 2695	50.01	170.5
5	Fe 2695	49.06	170.7
8	Fe 2695	41.51	171.9
3	Fe 82.2	48.10	170.4
5	Fe 82.2	49.14	169.9
8	Fe 82.2	36.03	171.00

Table 5.
Variation of the crystallinity and melting temperatures T_m with the content and sizes of powders.

As observed from the electrical conductivity measurements, the conductivity for the frequency ~2 × 10^4 Hz shows an increase with an order of magnitude, the highest values being at 8% concentration of Fe powders. For the investigated samples, the electrical conductivity increases (both in pure and in iPP with Fe powders) at increasing frequency. Increases of the loss factors were obtained with the increase of the metallic powder content, while the variations of the electrical permittivity were very small [30].

4.5.4 Incorporation of metallic powders into polymers

Incorporation of metallic powders into polymers is expected to impact the crystallization degree and the melting characteristics. DSC analyses were conducted to investigate the behavior of the studied materials. The melting temperatures and the crystallinity were determined from DSC thermograms (**Table 5**). The decrease of crystallinity with the amount of powders, if they are of the same type and are identical in size, was observed. This shows that the particles introduced into the iPP penetrate not only in the amorphous but also in the crystalline domains. The highest decrease was in all cases at 8% concentration of powder.

Crystallinity clearly depends on the size of the particles introduced: the larger the particles, the less they affect the crystallinity of the iPP. The decreases are highest in the case of small particles because they enter more easily in the crystalline domains. Interesting changes occur in melting temperatures. Generally melting temperature increases with the increase in powder content and at 8% concentration, temperature has the highest value.

4.5.5 SEM analysis

Figure 5(a and b) shows the morphological analyses for iPP with 8% concentration of powders with size of 82.2 nm, and **Figure 6(a and b)** shows the

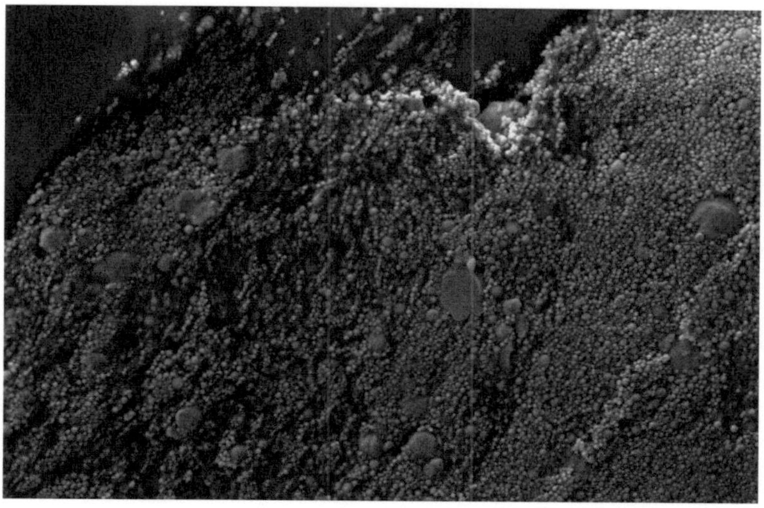

a

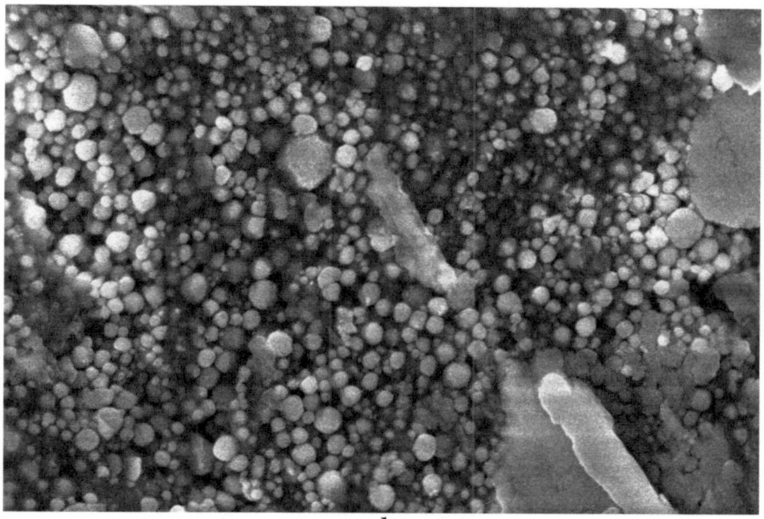

b

Figure 5.
SEM analysis of iPP with 8% concentration of Fe powders with size of 82.2 nm: 20,000× (a) and 50,000× (b).

morphological analysis for iPP with 8% concentration of powders with size of 2695 nm. A relatively uniform distribution of Fe particles with size of 82.2 nm is observed. They even entered the crystalline areas.

This confirms that extrusion preparation allows good dispersion of Fe particles in the PP matrix, a fact confirmed by other authors [18]. In this case the existence of some agglomerations of particles is observed. For the most part, these agglomerates are initiated by the voids resulting from partial detachment of the matrix.

Examination of **Figure 6** shows that the larger particles with size of 2695 nm penetrate first into the amorphous zone and then into the crystalline zone. This fact is in accordance with the crystallinity values obtained for these nanopolymers. In the figure agglomeration of particles also appears, especially in the amorphous areas or in areas with defects.

a

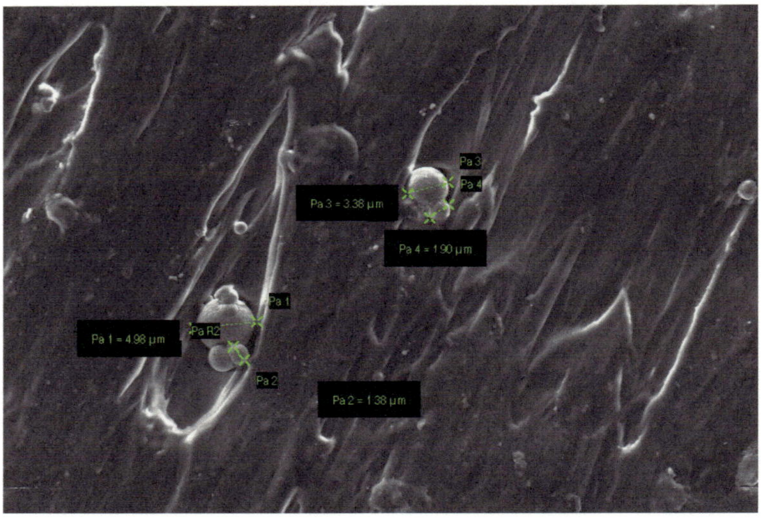

b

Figure 6.
SEM analysis of iPP with 8% concentration of Fe powders with size of 2695 nm: 1000× (a) and 5000× (b).

5. Conclusions

Bio PP materials for sutures show very good water resistance and present more important phenomenon of degradation in the presence of water for a longer period of time, over 40 days. Bio PP shows much more severe destruction of the crystalline domains in sodium solution in water. Bio PP meshes show the same degradation phenomena as those of bio PP sutures. For the samples that were introduced into water with beechwood, reduced degradation was obtained, variations comparable to those for samples introduced only in water.

An improvement in the elastic properties of the composite materials (iPP-Fe powder) was observed. In all cases studied, the increase in Shore hardness was achieved. The phenomenon of percolation has not been obtained. Thermal conductivity has little variations. It is noted that the highest Shore hardness values are obtained for composite iPP-Fe 82.8 nm, for 8% Fe concentration. This material also has the best thermal endurance. For applications in the field of biosensor construction, particles with dimensions of the order of nanometers are required.

Author details

Doina Elena Gavrila[1*], Victor Stoian[1], Alina Caramitu[2] and Sorina Mitrea[2]

1 Physics Department, University "Politehnica", Bucharest, Romania

2 National Institute for Research and Development in Electrical Engineering INCDIE, ICPE-CA, Bucharest, Romania

*Address all correspondence to: gavrila@physics.pub.ro

References

[1] Maddah HA. Polypropylene as a promising plastic a review. American Journal of Polymer Science. 2016;**6**(1): 1-11. DOI: 10.5923/jajps.20160601.01

[2] Harutun GK. Handbook of Polypropylene and Polypropylene Composites. 2nd ed. New York: Marcel Dekker; 2009. ISBN: 0-8247-4064-5

[3] Bostaca M, Gavrila DE. Studies on the use of polypropylene as alternative material to polyvinylchloride to improve human health and the state of environment. In: Abstracts 4th International Colloquium "Physics of Materials" PM4; Univ. "Politehnica" Bucharest, Romania. 2014. p. 102

[4] Bostaca M, Gavrila DE. Humidity in the composite material polyamide-polypropylene. In: Proceedings of DEIS, Conference on Solid Dielectrics; 30 June-4 July 2013; Bologna. Vol. 2. pp. 702-705

[5] Mohanty AK, Misra M, Drzal LT. Natural Fibers, Biopolymers and Biocomposites. London: CRC Press Taylor Francis; 2005. ISBN 0-8493-1741-X

[6] Meyers MA, Po-Yu C, Yu-Min Lin A, Seki Y. Biological materials: Structure and mechanical properties. Progress in Materials Science. 2008;**53**:1-206

[7] Chandra R, Rustgi R. Biodegradable polymers. Progress in Polymer Science. 1998;**23**:1273

[8] Hogan CB. Abiotic factor. In: Encyclopedia of Earth. Washington, D.C.: National Council for Science and the Environment; 2010

[9] Dunson WA. The role of abiotic factors in community organization. The American Naturalist. 1991;**138**(5):1067-1091. DOI: 10.1086/285270. JSTOR2462508

[10] Vasiliadou ES, Lemoniadou AA. Production of biopropylene using biomass-derived sources. In: Atwood A, editor. Sustainable Inorganic Chemistry. Chichester, West Sussex, UK: Wiley; 2016. pp. 129-139

[11] Gavrila DE, Stoica V. Analysis of degradation of polymer materials biopolypropylene and polyglycolide-caprolactone used in surgical sutures. In: Proceedings of the 7th IEEE International Conference on e-Health and Bioengineering (EHB 2019); 21-23 November 2019; Iasi, Romania: IEEE Xplore Database; www.cnas.ro>map>id

[12] Harloff J. Application of Polymers for Surgical Sutures (MSE 430) 17 April 1995. Department of Materials Science and Engineering: University of Pennsylvania; 1995

[13] Patel H, Ostergard DR, Sternschuss G. Polypropylene mesh and the host response. International Urogynecology Journal. 2012;**23**(6):660-679

[14] Iakovlev VV, Guelcher SA, Bendavid R. Degradation of polypropylene in vivo: A microscopic analysis of meshes explanted from the patient. Journal of Biomedical Materials Research Part B: Applied Biomaterials. 2017;**105**(2):237-248

[15] Wang J. Electrochemical biosensors: Towards point-of-care cancer diagnostics. Biosensors and Bioelectronics. 2006;**21**:1887-1892

[16] Hosu OA. New strategies in sensor design with analytical and bio analytical applications [thesis]. Cluj-Napoca: University of Medicine and Pharmacy; 2017

[17] Hanemann T, Vinga-Szabo D. Nanoparticles composites materials. Materials. 2010;**3**:3468-3517

[18] Kale C, Dhoka P, Goyal RK. Effect of processing route on electrical properties of polymer/iron composites. Journal of Electronic Materials. 2016;**45**(8):4148-4153

[19] Gungor A. The physical and mechanical properties of polymer composites filled with Fe powder. Journal of Applied Polymer Science. 2006;**99**(5):2438-2442. DOI: 10.1002/app.22637

[20] Lu C, Wang Q. Preparation of ultrafine PP/iron composite powders through pan milling. Journal of Materials Processing Technology. 2004;**145**(1):336-344

[21] Delogu F, Gorassi G, Sorrentino A. Fabrication of polymer nano composites via ball milling: Present status and future perspectives. Progress in Materials Science. 2017;**86**:75-126

[22] Tasdemir M, Ozkan Gulsoy H. Mechanical properties of polymers filled with iron powders. International Journal of Polymeric Materials. 2008;**57**(3):258-265. DOI: 10.1080/00914030701473636

[23] Bogoeva-Gaceva G, Raka L, Dimzoski B. Thermal stability of PP/organo-clay nanocomposites produced in single-step mixing procedure. Advanced Composites Letters. 2008;**17**(5):161-164. DOI: 10.1177/096369350801700503

[24] Caramitu AR, Mitrea S, Marinescu V, Ursan GA, Aradoaie M, Lingvay I. Dielectric behavior and morphostructural characteristics/metal nano powders. Materiale Plastice. 2019;**56**(1):103-109

[25] Gavrila DE, Caramitu A, Mitrea S, Paun C, Zburlea M. Effect of iron and aluminum powders on the properties of composites with isotactic polypropylene. In: Proceedings of the IEEE International Conference (ISFEE); 1-3 Nov. 2018. pp. 1-6. DOI: 10.1109/ISFEE.2018.8742420

[26] Bermann R. The thermal conductivity of dielectric solids at low temperature. Advances in Physics. 1953;**2**(5):103-140

[27] Nielsen LE. Thermal conductivity of particulate-filled polymers. Journal of Applied Polymer Science. 1973;**17**:3819-3820

[28] Mamunya YP, Davydenko VV, Pissis P, Lebedev EV. Electrical and thermal conductivity of polymers filled with metal powders. European Polymer Journal. 2002;**38**:1887-1897

[29] Krause B, Potschke P. Electrical and thermal conductivity of polypropylene filled with combinations of carbon fillers. In: Proceedings of the Regional Conference Graz. 2015. p. 040003

[30] Maharramov AM, Ramazanov MA, Sultanova JR, Hajiyeva FV, Hasanova UA. The structure and dielectric properties of nanocomposites based on isotactic polypropylene and iron nanoparticles. Journal of Optoelectronics and Biomedical Materials. 2016;**8**(3):113-118

Chapter 4

Synthesis and Characterization of PANI and Block Copolymer PANI-b-PEO Catalyzed by Maghnite (AlgerianMMT): Electrical and Electronic Domain

Abdelkader Rahmouni, Fatima Zohra Zeggai, Mohammed Belbachir, Bachari Khaldoun and Redouane Chebout

Abstract

Polyaniline (PANI) and its block copolymer ($PANI-PEO_{2000}$) has been prepared under effect of Maghnite-H+ (Algerian MMT) in different weight percentage (wt %) by cationic polymerization method. The structure of PANI and $PANI-PEO_{2000}$ is predicted by the FT-IR and ^{1}HNMR spectra. The thermal stability of homopolymer and block copolymer is confirmed by difference scanning calorimetry and analysis thermogravimetry. So after this results we can suggest that our heterogeneous catalyst called maghnite (Algerian MMT) can modified the morphology and the physical chemical properties of polyaniline (PANI) and its homolog block polyaniline-b-poly ethylene oxide ($PANI-b-PEO_{2000}$) in the mild conditions under microwave irradiation.

Keywords: ^{1}HNMR, green catalyst, green chemistry, conducting polymer, polyaniline, maghnite-H +, DSC, PEO

1. Introduction

Polyaniline (PANI) and its block copolymer (PANI-PEO) are the best promising material in conducting polymers, because of environmental stability, easy processing, and economical efficiency [1, 2]. PANI has been used for electrode of light emitting diode, Li ion rechargeable battery and corrosion protection [3, 4]. Nanocomposites (PANI-MMT) and (PANI-PEO-MMT) are interesting due to the special properties as abundance, low cost of MMT and attractive features such as a large surface area and ion- exchange properties [5, 6]. The clay is supplied by a local company known as ENOF Maghnia (Algeria) [7, 8]. Microwave heating has been found to be particularly advantageous for reactions under "dry" media [9, 10]. Microwave it's rapidly method in modern chemistry because offer a certain number of advantage, that it can be completed in a few seconds or minutes and without a solvent [2, 11]. Absence of solvent reduces the risk of explosions when reaction takes place in a microwave oven [12, 13]. The absence of solvent reduces the risk of explosions when the reaction takes place in a closed vessel in an oven [14, 15]. Aprotic dipolar solvents with high boiling points are expensive and difficult to remove from the reaction mixtures [16, 17]. The aim of this paper is to study the polymerization of aniline and its homolog block copolymer PANI-PEO catalyzed by Maghnite- H^+ under microwave irradiation [18]. This catalyst can be easily separated from the polymer product and regenerated by heating at a temperature above 100°C [19, 20]. The kinetics studies of different synthesis are discussed together with the mechanism of polymerization.

2. Experimental

2.1 Microwave apparatus

The temperature was maintained at 160°C in all experiments. Microwave irradiation was performed in a single mode focused CEM reactor (Model Discover, CEM Co., Matthew, NC) operating at 2.45 GHz with ability to control output power.

2.2 Materials and methods

MMT clay was obtained from ENOF Maghnia (Algeria). The MMT-H+ (Mag-H+) was prepared as described by Belbachir et al. and water (pH < 7) was used to synthesize emeraldine salt clay (PANI/Mag-H+) by cationic polymerization [16]. Polyethylene oxide (relative molecular mass of 2.10^3) was obtained from Sigma Aldrich. Polyaniline (PANI) homopolymer was prepared in laboratory of polymers chemistry (Oran University, Algeria) by standard chemical intercaled method [21].

2.3 Measurements

^{1}H nuclear magnetic resonance (NMR) measurements were carried out on a 300 MHz Bruker NMR Spectrometer equipped with a probe BB05 mm, in CDCl3. Fourier transform infrared spectroscopy (FTIR) spectra were obtained between 900 and 4000 cm -1 on an ATI Matson FTIR No 9501165. Intrinsic viscosity, [η], was measured at 30°C in benzene.GPC measurements of the samples were carried out using a WISP 712, Waters Associates chromatograph. The purification of polymers were carried out by dissolving the product in chloroform ($CHCl_3$) and filtering to eliminate the Maghnite-H+. Then, chloroform was removed by evaporation [22, 23].

2.4 Preparation of (Maghnite-H^+) as catalyst

Catalyst called (maghnite-H+) was prepared according to the process similar to that described by Belbachir et al. [24, 25]. The raw maghnite was placed in an erlenmeyer flask together with 100 ml of distilled water and a solution of sulfuric acid 0.25 M then stirred using a magnetic stirrer for 2 h at room temperature. After filtration up to pH 7, the activated maghnite-H+ is dried in the stove for 24 hours at 105°C for characterization [26].

2.5 Synthesis of polyaniline (PANI)

Into a flask with 100 mL and stirred to allow proper mixing was put a mixture of Maghnite-H^+ (5%) and solution of (H_2SO_4 0.25 M), adequate amount of aniline (0.05 mol) was added to a solution. The reaction mixture was then submitted to microwave irradiation at 160°C for 4 minutes. Finally, the mixture was cooled at room temperature, filtered and washed extensively with distilled water and methanol to remove catalyst and any unreacted aniline [27, 28].

2.6 Synthesis of block copolymer PANI-PEO

Copolymer PANI-PEO was synthesized via a cationic polymerization. To a (0.05 mol) of monomer aniline was added a solution of PEO (0.05 g in 25 mL of distilled water) and (5 wt %) of maghnite-H+ as an initiator. The mixture was stirred for 15 min. Then it is treated in a microwave oven at the power of 950 W, the temperature and viscosity of the reactive mixture increase fast and gelation point is reached after 4 min at 160°C. The mixture was cooled at room temperature, filtered and washed extensively with distilled water and methanol to remove impurity [16, 29].

3. Results and discussion

3.1 Spectroscopy characterization

Synthesized polyaniline was confirmed by noticing the predominant peaks at the wave numbers of 1501 cm^{-1}corresponding to C=C stretching of quinine ring, 1557 $cm^{-1}$for C=C stretching of benzenoid ring, 1293 $cm^{-1}$for C–N stretching, 755 and 838 $cm^{-1}$for C–H vibration of Para coupling benzenoid and benzene rings. Finally, C–H bending at 694–593 cm^{-1} corresponds to aromatic ring and 507 $cm^{-1}$is stretching at out of the plane [30, 31].

As shown in (**Figure 2**), there are significant changes in both the intensities and the frequencies in the product (PANI-PEO_{2000}). There are more pronounced between 690 and 1574 cm^{-1}, significant interaction between the oxygen of the ether group of PEO and the nitrogen in the PANI [32]. As shown in (**Figure 1**), polyethylene oxide (PEO_{2000}) show a band of methylene group (CH_2) stretching between 2950 and 2840 cm^{-1} and a large broad band appears centered at 3442 cm^{-1}which confirms that PEO_{2000} is highly hydrophilic [9].

Polyaniline was successfully synthesized as show in (**Figure 3**) and the different hydrogen peaks are present. The strongest sharp peaks centered at 7 and 7.8 ppm due to protons from phenylene and disubstituted phenylene units and the weak peak at 4.81 ppm due to (-NH) group but the peak at 6.22 ppm due to (–NH2) as end group [33, 34].

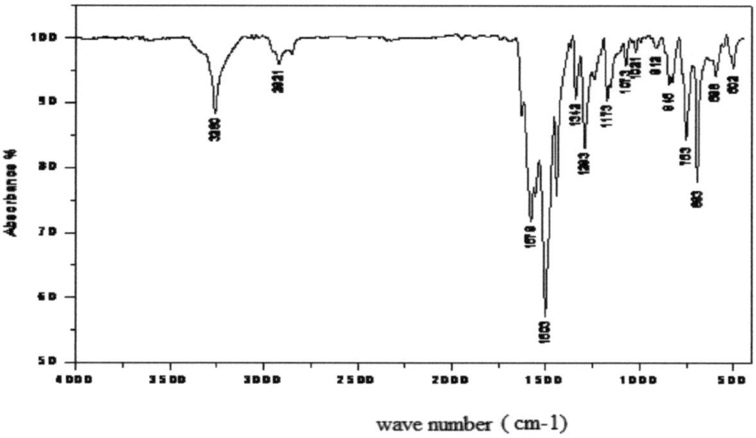

Figure 1.
FT-IR spectra of polyaniline (PANI) catalyzed by maghnite-H+ under microwave irradiation.

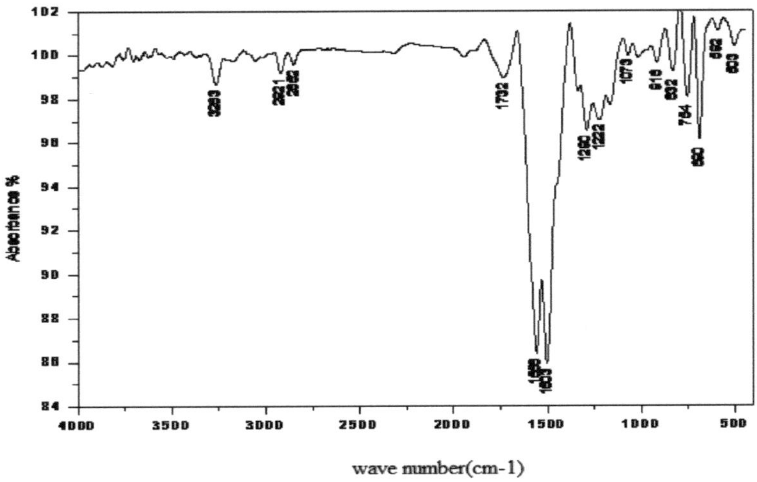

Figure 2.
FT-IR spectra of block copolymer (PANI-PEO$_{2000}$) catalyzed by maghnite-H+ under microwave irradiation.

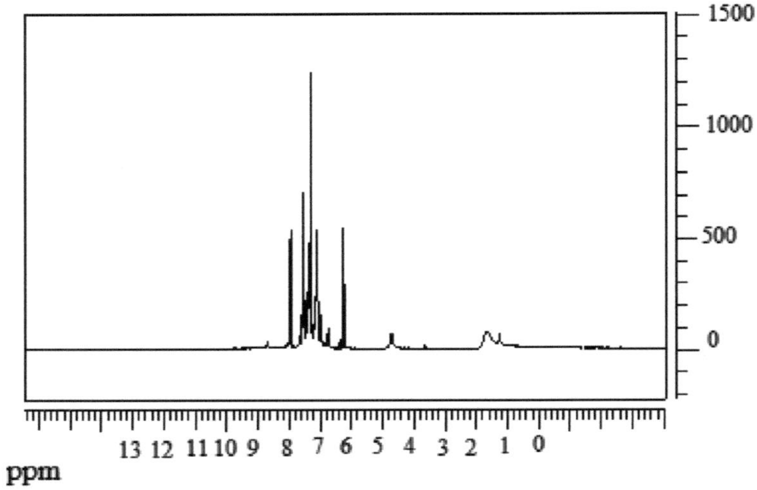

Figure 3.
Describes[1]H-NMR spectra of (PANI) catalyzed by maghnite-H+ under microwave irradiation.

^{1}H NMR spectroscopy at 300 MHz (Solvent $CDCl_3$) and according to the work published by Yahiaoui et al., (**Figure 4**) for pure PEO showed different peaks: (a) the methylene groups (CH_{2-}) at 2.6 ppm, and (b) the methylene (CH_2O^-) at 3.7 ppm [35, 36].

The block copolymer (PANI-PEO$_{2000}$) was confirmed by 1HNMR spectrum as show in (**Figure 5**). The wide signal in the region of 6.8 to 8 ppm was assigned to benzenoid hydrogen of polyaniline. Signals at 3.25–3.75 ppm indicate peak of CH_2O^- and $CH_2CH_2O^-$ hydrogen of polyethylene oxide reported [37]. Peak at 1.5–2 ppm is due to CH_2 hydrogen respectively [37, 38].

3.2 Optical properties

Conductive polymers synthesized PANI and PANI-PEO$_{2000}$ has a conjugated system of double bonds in a backbone polymer. The UV-visible spectral peak in the

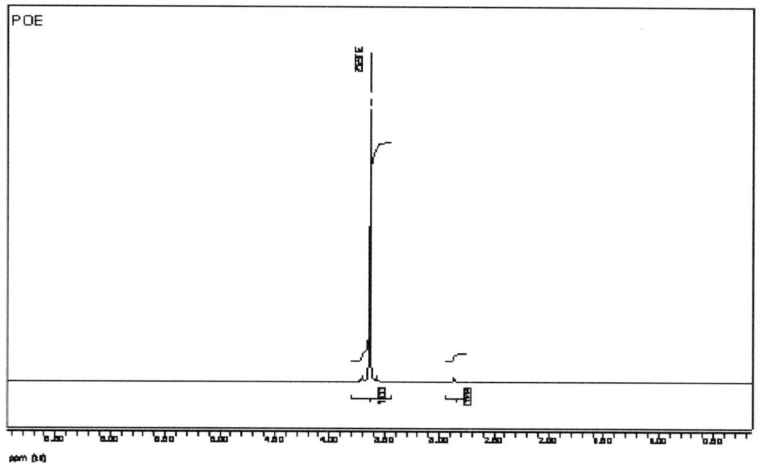

Figure 4.
Describes ^{1}H-NMR spectrum of poly ethylene oxide (PEO_{2000}).

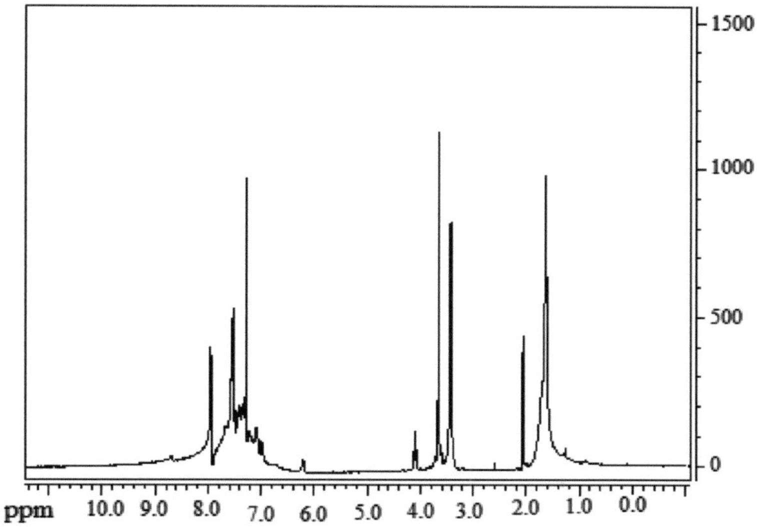

Figure 5.
Describes ^{1}H-NMR spectra of the copolymer (PANI-PEO_{2000}) catalyzed by maghnite-H+ under microwave irradiation.

250–300 nm region is due to the aniline groups and $\pi \rightarrow \pi^*$ is a conjugated couple system of the benzoic states in the 350 to 400 nm regions (**Figure 6**) [39, 40].

3.3 Thermal properties

As shown in (**Figures 7** and **8**), the thermogram analysis of catalyst (maghnite-H+) shown two stages of weight loss. The weight loss in below 100°C is a result of free water and the weight loss around 600°C is associated with the dehydroxylation of silicate structure [37, 41].

For the thermogram analysis (TGA) of polyaniline (PANI), it can be found that the weight loss amounted 61, 17% at the temperature range of 187–600°C, which be

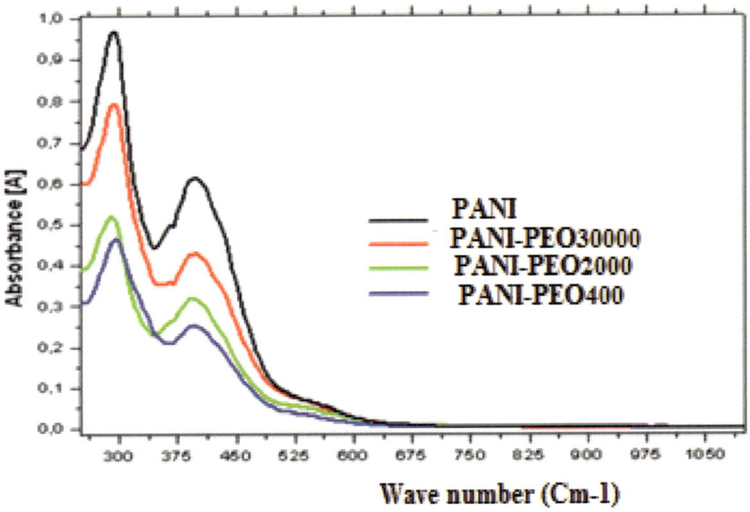

Figure 6.
UV spectral of the different form of (PANI-PEOs) catalyzed by maghnite-H+ under microwave irradiation.

true because polyaniline it is known a hygroscopic polymer as show in (**Figure 9**) [42, 43]. in the DSC thermogram of the PANI as show in (**Figure 10**), there were two endothermic peaks at 55.99 and 103.46°C.Therefore, these endothermic peaks were due to the evaporation of water, this is in agreement with the literature [44, 45]. The glass transition (Tg) appears at 74.06°C [46].

The curves of weight loss versus temperature showing the behavior of PANI-PEO_{2000} sample was presented in (**Figure 11**). The first significant weight loss occurs already at temperature between 50 and 100°C, that PANI-PEO_{2000} is hygroscopic and during the heating to 100°C the residual water evaporates [46]. Then the main mass loss, which corresponds to polymer degradation starts at about 200 and 500°C [47].

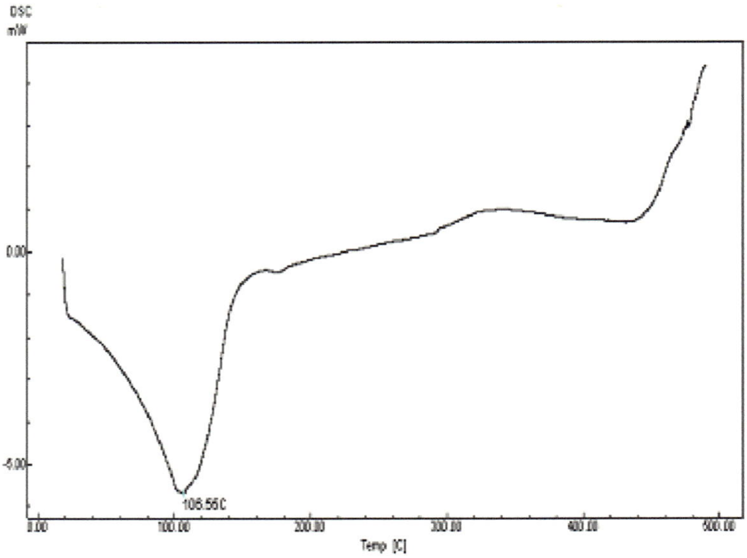

Figure 7.
Thermogram analysis measurements (DSC) of maghnite-H+ (heating rate 10°C/mn) catalyzed by maghnite-H+ under microwave irradiation.

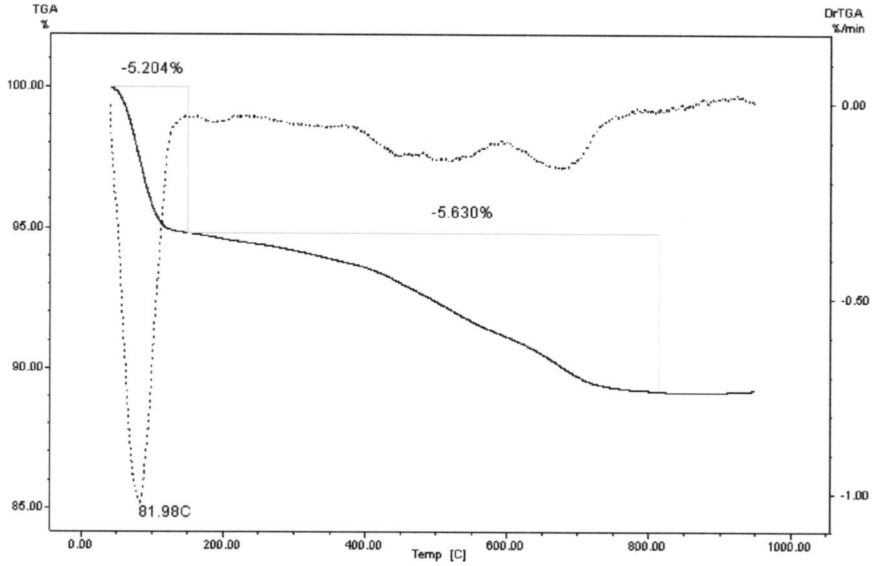

Figure 8.
Thermogram analysis (TGA) curves of a maghnite-H⁺ obtained in nitrogen atmosphere at heating rate of 10°C/min.

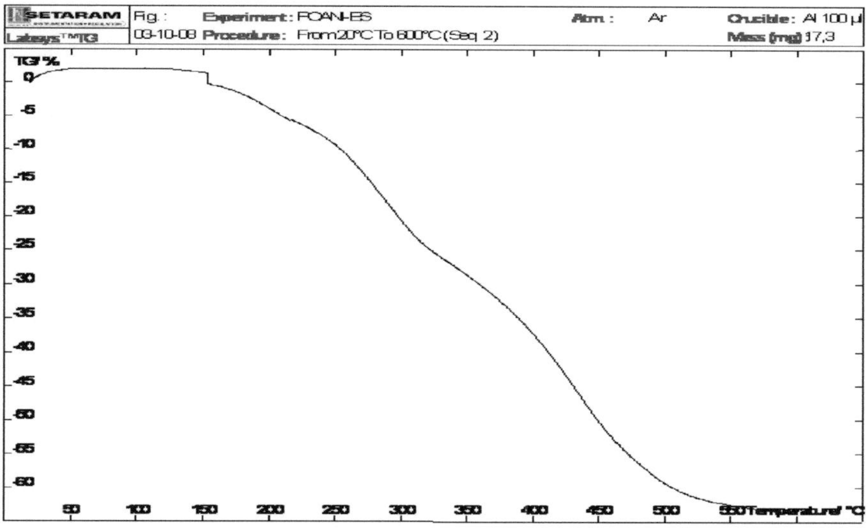

Figure 9.
Thermogram analysis (TGA) curves of PANI prepared in the presence of Maghnite-H+ (0.25 M) under microwave irradiation.

In polyaniline (PANI) monomer and block copolymer (PANI-PEO2000) as show in (**Figure 12**). Firstly, we notice the presence of two endothermic peak at (68, 39 and 190, 09°C) it is associated respectively to the evaporation of water absorbed by the copolymer and melting POE2000 block [48, 49]. The glass transition temperature of block copolymer (Tg = 16, 79°C) which is in agreement with the literature [50].

3.4 Gel permeation chromatography

Table 1 describes thermodynamics properties of PANI and its block PANI-PEO_{2000}. **Tables 2** and **3** describes the molecular weight distribution averages for the polymer

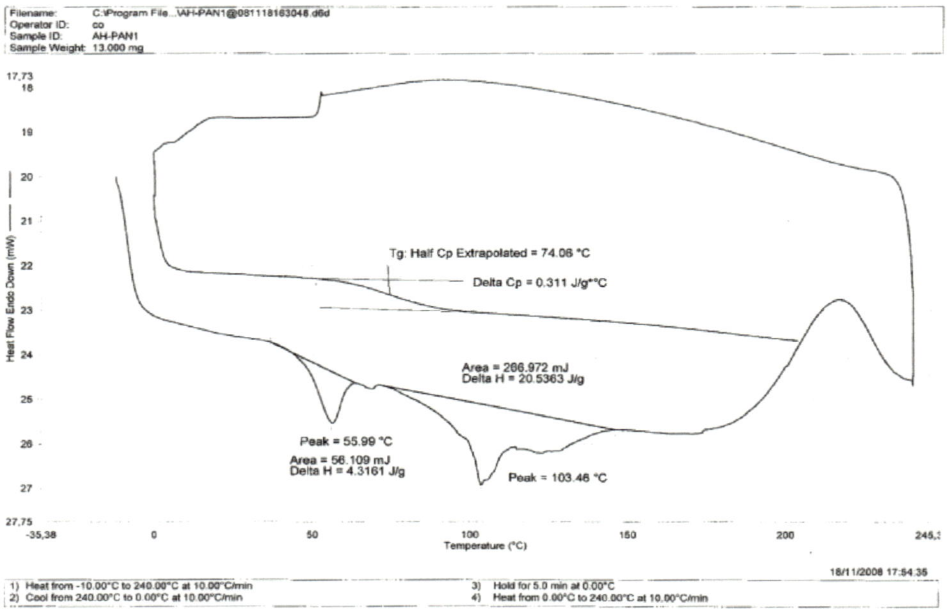

Figure 10.
Describes thermogram analysis (DSC) of PANI (heating rate 10°C/mn) catalyzed by maghnite-H+ under microwave irradiation.

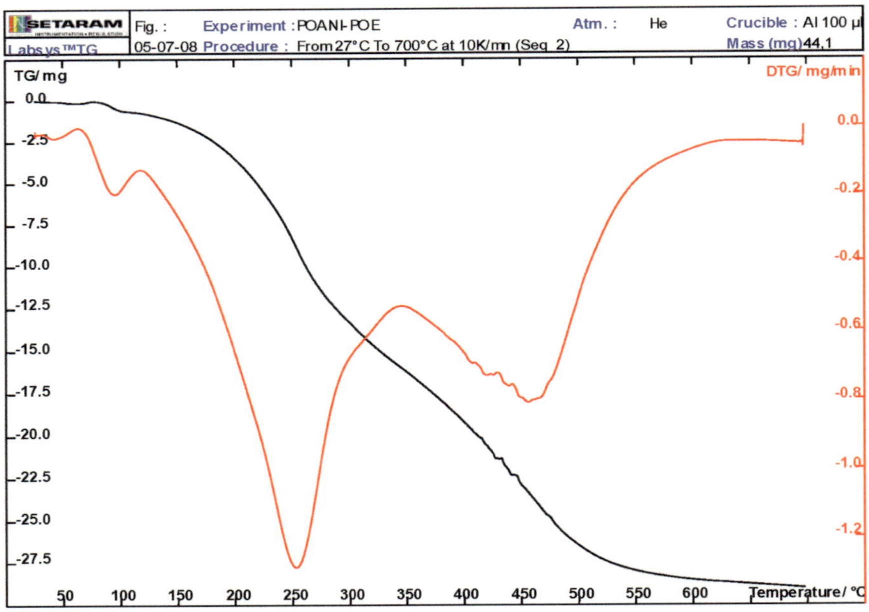

Figure 11.
Thermogram analysis (TGA) curves of PANI-PEO$_{2000}$ prepared in the presence of maghnite-H+ (0.25 M) under micro wave irradiation.

and it's copolymer in the other hand (**Figures 13** and **14**) indicate a bimodal distribution. The macromolecular weight distribution of the obtained polymer and copolymer are narrow, this confirming the formation of the polymer PANI and the block copolymer PANI-PEO [51, 52].

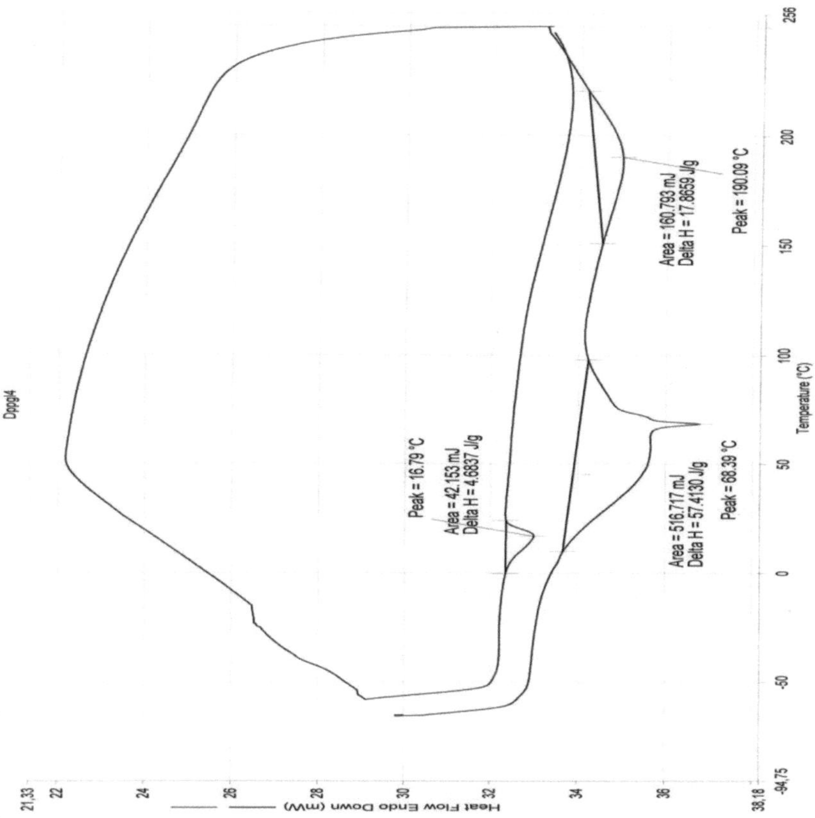

Figure 12.
Thermogram analysis (DSC) of PANI-PEO$_{2000}$ (heating rate 10°C/mn) catalyzed by maghnite-H+ under microwave irradiation.

Sample	T_1	T_2	T_3	T_g	ΔH_1	ΔH_2	ΔH_3	ΔC_p
PANI	55.99	03.46	X	74.06	4.3161	20.5363	X	0.311
PANI-PEO$_{2000}$	68.39	190.09	X	16.79	57.4130	17.8659	X	4.6837

Table 1.
Describes thermodynamics properties of PANI and its block PANI-PEO$_{2000}$.

Sample name	RT	Area	% Area	Mn	Mw	Polydispersity
PANI	17.973	1,318,425	6.01	644	746	1.15
	23.638	20,616,195	93.99	33	114	3.41

Table 2.
Gel permeation chromatograph of PANI composite in THF catalyzed by maghnite-H^+ under microwave irradiation.

3.5 Electrical properties

The value of transverse strength and the electrical conductivity of the PANI and its block copolymer PANI-PEO2000 were calculated from Eqs. (1) and (2) as show in (**Table 4**) [53].

Sample name	RT	Area	% Area	Mn	Mw	Polydispersity
PANI-PEO2000	17.917	130,748	1.30	2664	2701	1.013896
	23.032	9,960,549	98.70	50	264	5.342141

Table 3.
Gel permeation chromatograph of block copolymer PANI-PEO$_{2000}$ composite in THF catalyzed by maghnite-H^{+} under microwave irradiation.

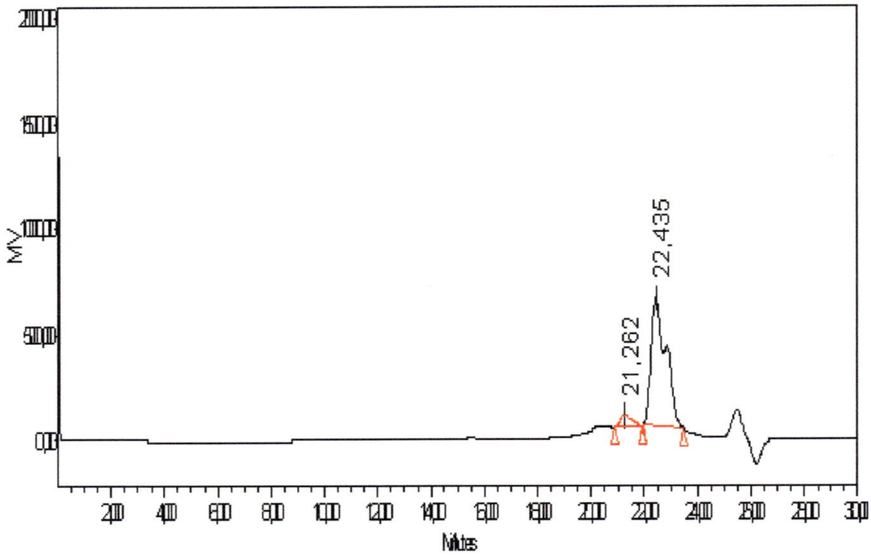

Figure 13.
Describes gel permeation chromatograph of PANI with 2% maghnite-H^{+} at 160°C for 4 min.

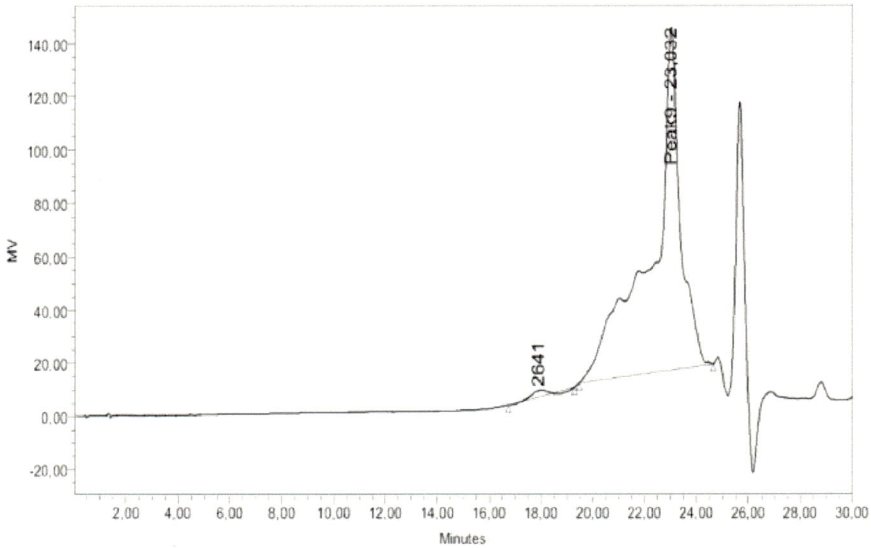

Figure 14.
Describes gel permeation chromatograph of PANI-PEO$_{2000}$ with 2% maghnite-H^{+} at 160°C for 4 min.

$$\rho = R\ (\pi.r^2 / e)\ (\Omega.cm) \tag{1}$$

$$\sigma = 1/\rho (S/cm) \tag{2}$$

3.6 Solubility of product

The polyaniline powder was added to 50 ml of different solvent (DMF, Acetonitrile, toluene, dichloromethane, THF, and chloroform). The dry weight of the filter paper was used to calculate the solubility of the composites. The best solvents for PANI and its block are determined to be DMF and Toluene as show in (**Table 5**) and (**Figure 15**). Finally, we can calculate the band of energy by equation (3) [54].

Sample	e (cm)	R(Ω)	ρ (Ω. cm)	σ (S/cm)
PANI	0.1	1.880	25.792	0.038
PANI-PEO_{2000}	0.1	0.976	0.163	6.134

Table 4.
Describes electrical properties of PANI and PANI-PEO catalyzed by maghnite-H^+ under microwave irradiation.

Solvant	PANI	PANI-PEO_{2000}
DMF	0.55	1.6
Toluene	0.52	1.4
Chloroforme	0.47	1.05
Dichloromethane	0.39	0.95
Acetonitrile	0.20	0.79
THF	0.12	0.58

Table 5.
Describes solubility parameters of PANI and PANI-PEO_{2000} composites catalyzed by maghnite-H^+ under microwave irradiation in different solvents (g/ml).

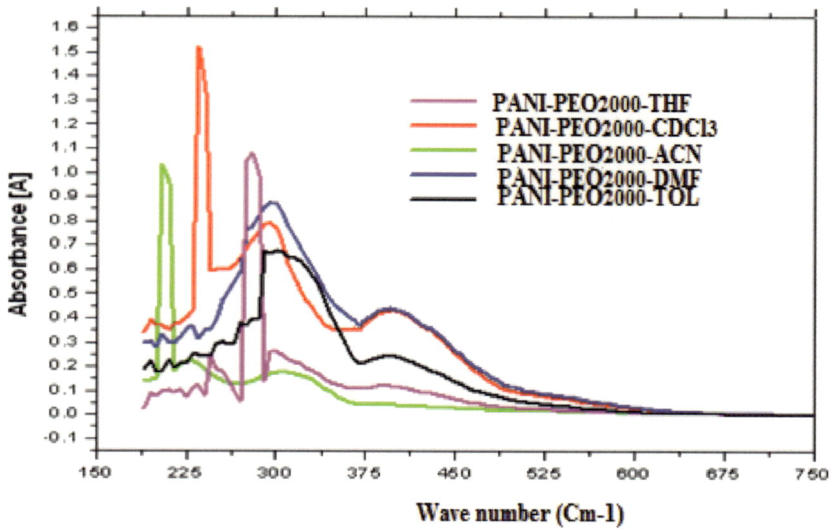

Figure 15.
Describes UV spectral of the copolymer (PANI-PEO_{2000}) catalyzed by maghnite-H+ under microwave irradiation in different organic solvents.

Concentration	C	C_1=2C/3	C_2=C/2	C_3= C/3	C_4=C/4
Average	3.60	3.61	3.61	3.60	3.59
Cinema (cst)	0.61	0.61	0.61	0.60	0.6
Dynamic (cp)	0.91	0.90	0.92	0.89	0.89
Relative	1.00	1.01	1.01	1.01	1
Specific	0.00	0.01	0.00	0.00	0
Reduced	12.22	13.99	19.45	20.96	22.40
Inherent	12.10	13.90	19.45	20.90	22.41
Solomon	12.14	13.90	19.50	20.95	22.43
Visct. Intr (ml/g^{-1})	25.80		25.78		25.77

Table 6.
Describes viscosimertic properties of pure PANI catalyzed by maghnite-H+ under microwave irradiation.

Concentration	C	C_1=2C/3	C_2=C/2	C_3= C/3	C_4=C/4
Average	3.61	3.61	3.60	3.61	3.58
Cinema (cst)	0.61	0.60	0.61	0.61	0.61
Dynamic (cp)	0.91	0.91	0.91	0.88	0.90
Relative	1.00	1.00	1.01	1.01	1.01
Specific	0.00	0.01	0.01	0.01	0.00
Reduced	7.49	12.10	14.37	13.55	18.05
Inherent	7.40	12.05	14.31	13.50	17.99
Solomon	7.44	12.07	14.33	13.51	18.01
Visct. Intr (ml/g^{-1})	19.75		19.66		19.68

Table 7.
Describes viscosimertic properties of PANI-PEO$_{2000}$ catalyzed by maghnite-H+ under microwave irradiation.

$$\Delta E = h\nu = h\, c\, / \lambda \tag{3}$$

The solubility parameter (δ) is usually calculate by equation (4) and expressed in (cal/cm^3)$^{1/2}$ or preferably (j/cm^3)$^{1/2}$ units for many compound is defined from Hildebrand-Scotchard Solution theory is:

$$\delta = (\Delta E / V)^{1/2} \tag{4}$$

3.7 Viscosimertic parameters of PANI and PANI-PEO

Tables 6 and **7** describe intrinsic viscosity and properties physics of (PANI) and block (PANI-PEO$_{2000}$) according to the Mark-Houwink equation:

$$[\eta]\ (\mathrm{mg.l^{-1}}) = K\ [M_v]^a$$

4. Kinetics studies

4.1 Effect of the amount of mag-H + on the yield of copolymerization

Figure 9 describe the effect of the amount of catalyst on the yield of this copolymerization (PANI-PEO). As can be seen in (**Figure 16** and **19**), the copolymerization rate increased with the amount of Mag-H + and reaches a maximum at 160°C with (5% wt) of catalyst, above this temperature and percentage of catalyst, the yield decreases.

The increase in yield with temperature and molecular weight of catalyst is mainly due to the number of active sites in the catalyst responsible for initiating the reaction. Similar results are obtained by many research [55, 56].

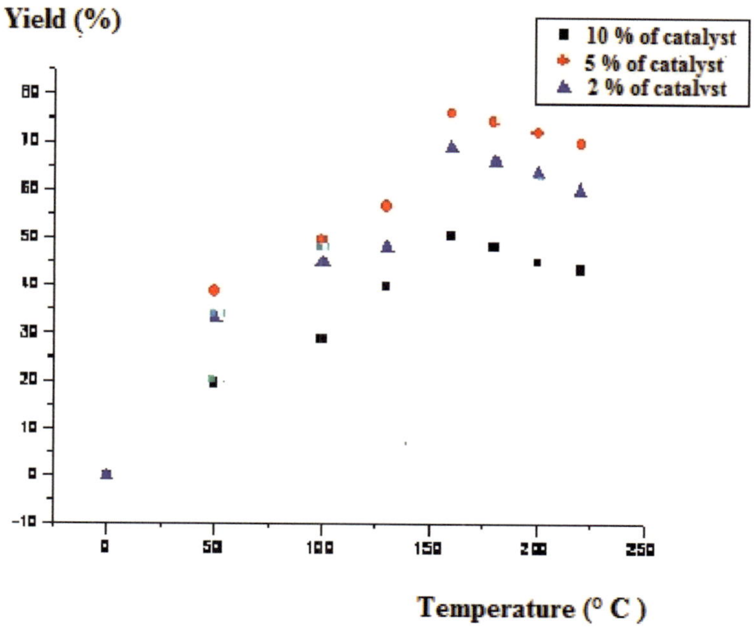

Figure 16.
Describes effect of the amount of mag-H+ (catalyst) on the yield of copolymerization.

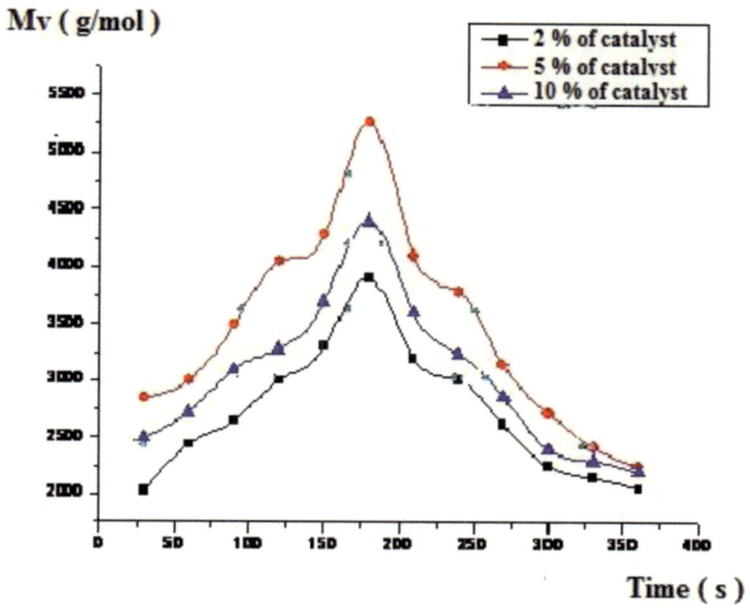

Figure 17.
Describes effect of the amount of mag-H + on the viscosimertic molecular weight of copolymerization.

4.2 Effect of the amount of Mag-H + on the molecular weight of copolymer

We have used (2%,5% and 10%) by weight as the amount maghnite and varying time after keeping the other parameters (the amount of monomers and the temperature). It is observed that the average molecular weight increases with time and reaches a maximum at 4 minutes of reaction and decrease after this time as show in (**Figure 17**) [57].

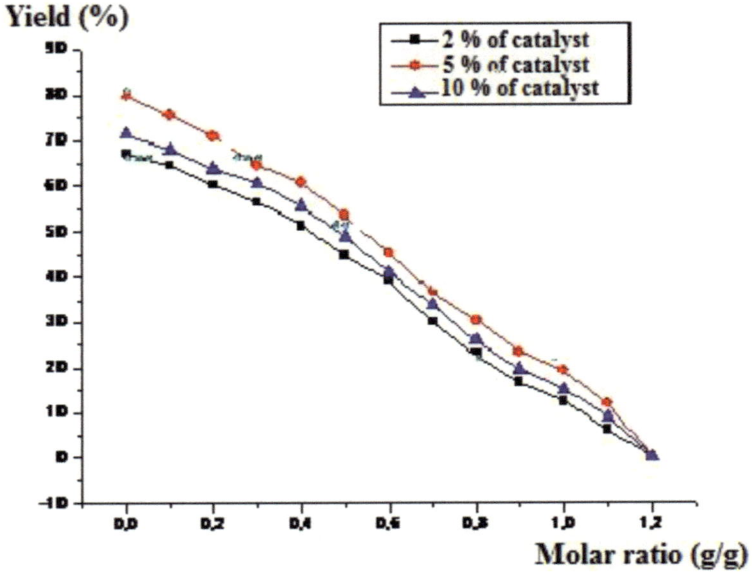

Figure 18.
Describes the effect of the molar ratio(ANI/PEO) upon the yield of copolymerization.

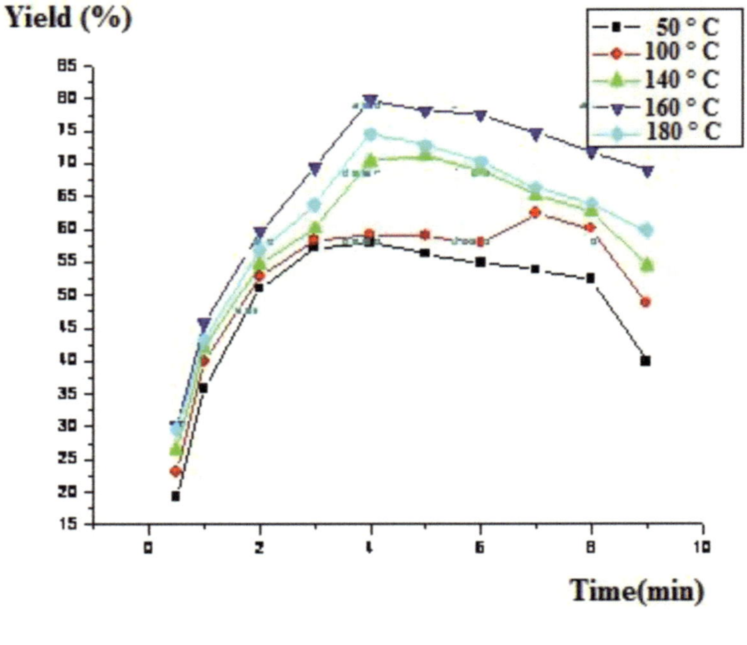

Figure 19.
Describes effect of temperature on the yield of copolymerization.

4.3 Effect of molar ratio aniline/PEO on the yield of copolymerization

We notice a significant change in the yield with increasing the amount ratio, in particular with increasing the amount of PEO used in this reaction processing. The (**Figure 18.**) below summarizes the influence of the molar ratio on Aniline/PEO upon

1-Initiation

The reaction occurs within the layers of the Maghnite exchanged by H^+ protons, which are capable of initiating cationic polymerization.

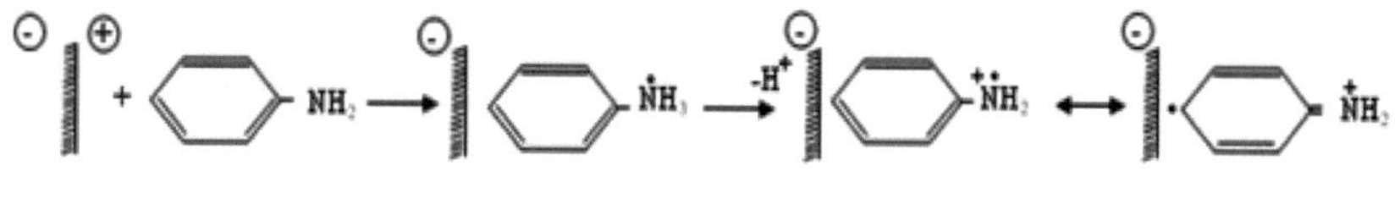

2-Propagation

The successive addition of monomers on the chain macromere growing present propagation process.

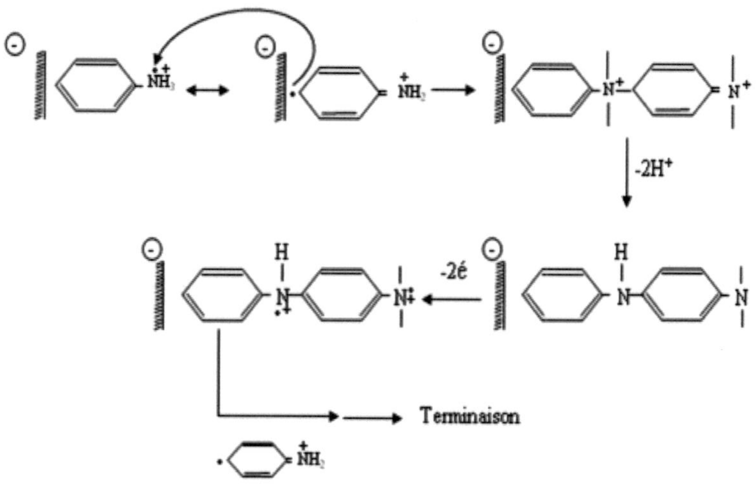

3-Termination

The termination takes place during the recombination of two radical ions giving rise to the final polymer.

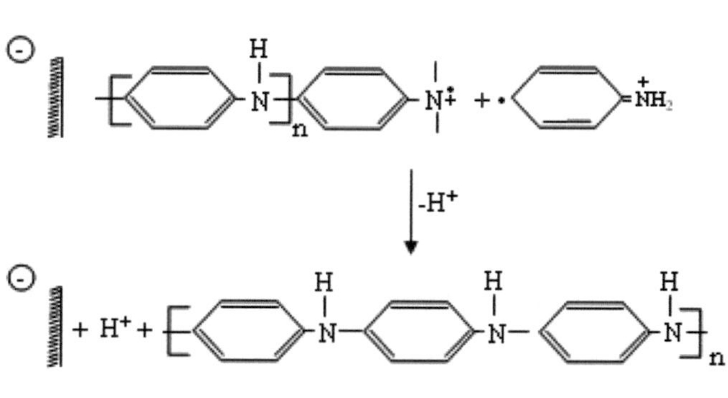

Figure 20.
Proposed mechanism of homopolymer (PANI) catalyzed by Maghnite-H + under microwave irradiation.

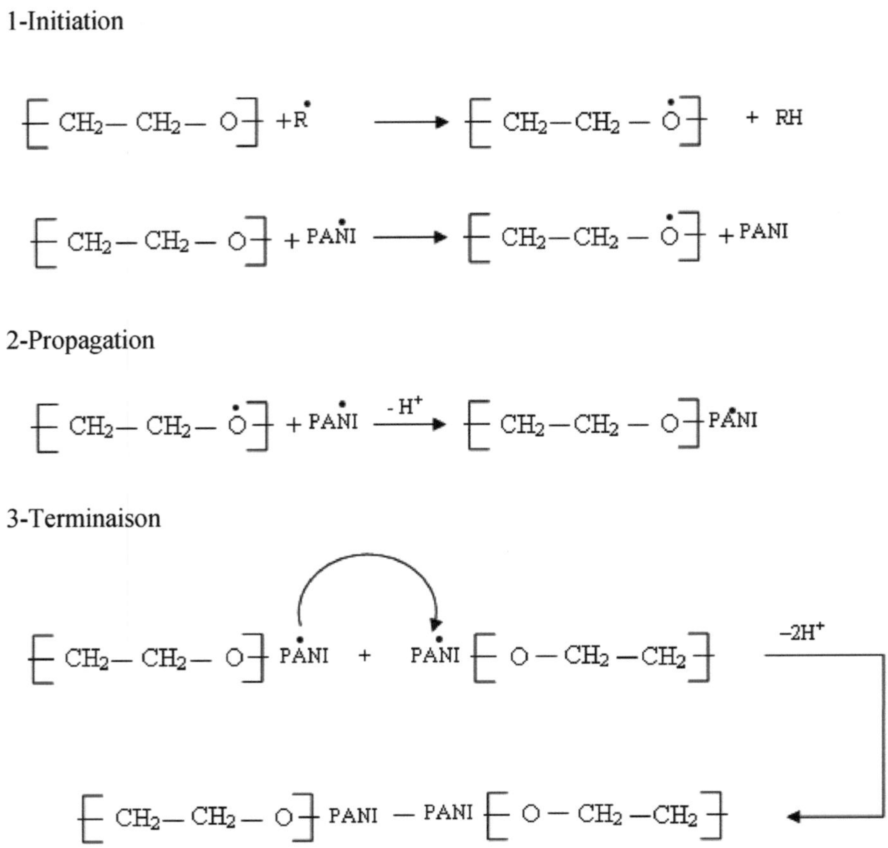

Figure 21.
Proposed mechanism of block copolymer (PANI-PEO$_{2000}$) catalyzed by Maghnite-H+ under microwave irradiation.

the reaction yield. This phenomenon can explain by the high reactivity and solubility of the POE in water compared to the aniline in particular at high temperature [58].

5. Proposed mechanism of polymerization and copolymerization

As shows in (**Figures 20** and **21**), the process of synthesis of PANI-Maghnite-H$^+$ and its homolog PANI-PEO$_{2000}$-Maghnite-H+ composites can be divided into the following three steps [59–61].

6. Conclusion

Based on Maghnite-H$^+$ (Algerian ecologic catalyst MMT) and under microwave irradiation our polymer (PANI) and block copolymer (PANI-b-PEO) were successfully synthesized and investigate. This product was prepared in order to combine the mechanical and physical properties of PEO$_{2000}$ with conducting properties of PANI. A possible mechanism of this cationic polymerization is discussed based on the results of the ^{1}H NMR Spectroscopic analysis of these compounds. Thus all the two types of composites (PANI and PEO) provide opportunities and rewards creating new world wide interest in these new materials in electronics devises.

Acknowledgements

All work was supported by the DGRSDT of Algeria republic and was performed using the equipment of the center of research scientific and technics in analysis chemical and physical (CRAPC)-Tipaza-Algeria.

Author details

Abdelkader Rahmouni[1*], Fatima Zohra Zeggai[1,2], Mohammed Belbachir[1], Bachari Khaldoun[2] and Redouane Chebout[2]

1 Department of Chemistry, Laboratory of Polymer Chemistry Faculty of sciences, Oran1 University Ahmed Benbella, Oran, Algeria

2 Centre de Recherche Scientifique et Technique en Analyses Physico-Chimiques (CRAPC), Tipaza, Algeria

*Address all correspondence to: ramaek23@yahoo.fr

References

[1] Carini GD, Angelo G, Tripodo G, Bartolotta A, Di Marco G. Low temperature excess specific heat and fragility in semicrystalline polymers. Physical Review B. 1996;**54**:15056-15063

[2] Brune DA, Bicerano J. Micromechanics of nanocomposites comparison of tensile and compressive elastic moduli, and prediction of effects of incomplete exfoliation and imperfect alignment on modulus. Polymer. 2002;**43**(2):369-387

[3] Mac Diarmid A. Synthetic metals: a novel role for organic polymers. Angewandte Chemie, International Edition. 2001;**40**:2581-2590

[4] Rajesh AT, Kumar D. Recent progress in the development of nano-structured conducting polymers/nanocomposites for sensor applications. Sensors and Actuators B: Chemical. 2009;**136**:275-286

[5] Zhang LJ, Long YZ, Chen ZJ, Wan MX. The effect of hydrogen bonding on self assembled polyaniline nanostructures. Advanced Functional Materials. 2004;**14**:693-698

[6] Park S, Chung SW, Mirkin CA. Hybrid organic-inorganic,rod-shaped nanoresistors and diodes. Journal of the American Chemical Society. 2004;**126**:11772-11773

[7] Jaroslav S, Irina S, Miroslava T. Polyaniline nanostructures and the role of aniline oligomers in their formation. Progress in Polymer Science. 2010;**35**:1420

[8] Zhang XY, Manohar SK. Polyaniline nanofibers: Chemical synthesis using surfactants. Chemical Communications. pubs.rsc.org. 2004:2360-2361

[9] Ray S, Okamoto M. Polymer/layered silicate nanocomposites: A review from preparation to processing. Progress in Polymer Science. 2003;**28**(11):1539-1641

[10] Rahmouni A, Harrane A, Belbachir M. 1H-NMR spectra of conductive, anticorrosive and Soluble polyaniline exchanged by an eco-catalyst layered (Maghnite-H+). wod. Journal of Chemistry. 2013;**8**(1):20-26

[11] Jang J. Conducting polymer nonmaterials and their applications. Advances in Polymer Science. 2006;**199**:189-259

[12] Li X, Tian SJ, Ping Y, Kim DH, Knoll W. One-step route to the fabrication of highly porous polyaniline nanofiber films by using PS-b-PVP diblock copolymers. As templates. Langmuir. 2005;**21**:9393-9397

[13] Long YZ, Duvail JL, Wang QT, Li MM, Gu CZ. Electronic transport through crossed conducting polymer nanowires. Journal of Materials Research. 2009;**24**:3018-3022

[14] Johnson SD, Anderson JM, Marchant RE. Biocompatibility studies on plasma polymerized interface materials encompassing both hydrophobic and hydrophilic surfaces. Journal of Biomedical Materials Research. 1992;**26**:915-935

[15] Raghunathan A, Kahol PK, Ho JC, Chen YY, Yao YD, Lin YS, et al. Low temperature heat capacities of polyaniline and polyaniline polymethylmethacrylate blends. Physical Review B 1998;**58**:15955-15958

[16] Belbachir M, Bensaoula A. Composition and Method for Catalysis Using Bentonites. United States Patent Number: 6274527 B1. 2001

[17] Belbachir M, Yahiaoui A, Hachemaoui A. An acid exchanged montmorillonite clay-catalyzed

synthesis of polyepichlorhydrin. International Journal of Molecular Sciences. 2003;**4**:548-561

[18] Kumar KV, Rao UVS. Thin solid electrolyte systems based on PEO and their application as an electrochemical cell. In: Proceedings of 2nd International Conference on Ionic Devices, 28-30th November. Chennai, India; 2003

[19] Choi HJ, Kim JW, Joo J, Kim BH. Synthesis and electrorheology of emulsion intercalated PANI-clay nanocomposite. Synthetic Metals. 2001;**121**(1-3):1325-1326

[20] Kuilla T, Bhadra S, Yao DH, Kim NH, Bose S, Lee JH. Recent advances in graphene based polymer composites. Progress in Polymer Science. 2010;**35**:1350-1375

[21] Spitalsky Z, Tasis D, Papagelis K, Galiotis C. Carbon nanotube polymer composites: Chemistry,processing, mechanical and electrical properties. Progress in Polymer Science. 2010;**35**:357-401

[22] Shin MK, Kim YJ, Kim SI, Kim SK, Lee H, Spinks GM, et al. Enhanced conductivity of aligned PANi/PEO/ MWNT nanofibers by electro-spinning. Sensors and Actuators B: Chemical. 2008;**134**:122-126

[23] Pinto NJ, Johnson AT, Mueller CH, Theofylaktos N, Robinson DC, Miranda FA. Electrospun polyaniline/ polyethylene oxide nanofiber field-effect transistor. Applied Physics Letters. 2003;**83**:4244-4246

[24] Yang R, Shichao Z, Lan Z, Wenbo L. Electrical properties of composite polymer electrolytes based on PEO-SN-LiCF3SO3. International Journal of Electrochemical Science. 2013;**8**:10163

[25] Devindrappa HU, Subba R, Ambika P. Study of dc conductivity and battery application of PEO/PANI Composites. Journal of Power Sources. 2006;**155**:3689

[26] Kunteppa H et al. AC conductivity and battery application of polyethylene oxide/PANI/ sodium chlorate composites. Advanced Materials Letters. 2013;**11**:856

[27] Gray KB, Rodrigues SJ. Poly (ethyleneoxide)-based composite electrolytes: Crystalline amorphous transition. Journal of the Electrochemical Society. 2001;**148**:1336

[28] Sujeet KC, Rajendra KS. Electrical conductivity studies on composite polymer electrolyte based on ionic liquid. Phase Transitions. 2010;**83**:457

[29] Rahmouni A, Harrane A, Belbachir M. Thermally stable forms of pure polyaniline catalyzed by an acid-exchanged MMT clay called maghnite-H+ as an effective catalyst. International Journal of Polymer Science. 2012;**2012**:846710, 7 p. DOI: 10.1155/2012/846710,2012

[30] Belbachir M, Bensaoula A. Composition and method for catalysis using bentonites. US Patent No. 7, 094-823 B2. 2006

[31] Megherbi A, Meghabar R, Belbachir M. Preparation and characterization of clay (Maghnite-H)/ poly (3,4-ethylenedioxythiophene) composites. Journal of Surface Engineering Materials and Advanced Technology. 2013;**3**:21-27

[32] Olad A, Rashidzadeh A. Preparation and anticorrosive properties of PANI/Na-MMT and PANI/OMMT nanocomposites. Progress in Organic Coating. 2008;**622**:93

[33] Norris ID, Shaker MM, Ko FK, Macdiarmid AG. Electrostatic fabrication of ultrafine conducting fibers: Polyaniline/polyethylene

oxide blends. Synthetic Metals. 2000;**114**(2):109-114

[34] Yahiaoui A, Hachemaoui A, Belbachir M. Synthesis and structural analyses of poly (1, 2-cyclohexene oxide) over solid acid catalyst. International Journal of Molecular Sciences. 2003;**4**:548-561

[35] Yahiaoui A, Belbachir M, Hachemaoui A. International Journal of Molecular Sciences. 2003;**4**:548-561

[36] Li Q, Cruz L, Phillips P. Granular-rod model for electronic conduction in polyaniline. Physical Review B. 1993;**47**:1840-1845

[37] Wright PV. Electrical conductivity in ionic complexes of poly(ethylene oxide). British Polymer Journa. 1975;**7**:319

[38] Kherroub DE, Belbachir M, Lamouri S. Nylon 6/clay nanocomposites prepared with Algerian modified clay (12-maghnite). Research on Chemical Intermediates. 2015;**41**:5217-5228

[39] Nascimento GM, Temperini ML. Structure of polyaniline formed in different inorganic porous materials: A spectroscopic study. European Polymer Journal. 2008;**44**:3501

[40] Zhang L, Zujovic ZD, Peng H, Bowmaker GA, Kilmartin PA, Travas-Sejdic J. Structural characteristics of polyaniline nanotubes synthesized from different buffer solutions. Macromolecules. 2008;**41**:8877-8884

[41] Rahmouni A, Harrane A, Belbachir M. Kinetics study of catio-radically polymerization of aniline catalyzed by Maghnite-Na clay catalyst layered (Western Algeria). World Journal of Nano Science & Technology. 2012;**1**(2):26-32

[42] Ayat M, Rahmouni A, Belbachir M. Methyl methacrylate and alpha-methyl styrene: New strategy for synthesis of bloc copolymers for use in potential biomedical applications generated by an ecologic catalyst called maghnite (Algerian MMT). Bulletin of Chemical Reaction Engineering and Catalysis. 2016;**11**:316-329

[43] Rahmouni A, Belbachir M. Green synthesis of cationic polyacrylamide composite catalyzed by an ecologically catalyst clay called maghnite-H+ (Algerian MMT) under microwave irradiation. Bulletin of Chemical Reaction Engineering and Catalysis. 2016;**11**:170-175

[44] Kao HM, Tsai YY, Chao SW. Functionalizedmesoporoussilica MCM-41 in poly (ethyleneoxide)-basedpolymerelectrolytes: NMR and conductivitystudy. Solid State Ionics. 2005;**176**:1261-1270

[45] Xia JY, Qiua XP, Ma XM, Cui MZ, Yang J, Tang XZ, et al. Composite polymer electrolyte doped with mesoporous silica SBA-15 for lithium polymerbattery. Solid State Ionics. 2005;**176**:1249-1260

[46] Xi JY, Tang XJ. Enhanced lithium ion transferencenumber and ionicconductivity of composite polymerelectrolytedopedwithorganic–inorganichybrid P123@SBA-15. Chemical Physics Letters. 2004;**400**:68-73

[47] Tominaga Y, Hong IC, Asai S, Sumita M. Proton conduction in Nafion composite membranes filledwithmesoporoussilica. Journal of Power Sources. 2007;**171**:530-534

[48] Gao Q, Xu Y, Wu D, Sun Y, Li XA. PH-responsive drug release frompolymer-coated mesoporous silica spheres. Journal of Physical Chemistry C. 2009;**113**:12753-12758

[49] Chu HQ, Yu C, Wan Y, Zhao DY. Synthesis of ordered mesoporous

bifunctional TiO2- SiO2-polymer nanocomposites. Journal of Materials Chemistry. 2009;**19**:8610-8618

[50] Praveen S, Sun ZF, Xu JG, Patel A, Wei Y, Ranade R, et al. Compression and agingproperties of experimental dental composites containingmesoporoussilica as fillers. Molecular Crystals and Liquid Crystals. 2006;**448**:223-231

[51] Jiao J, Sun X, Pinnavaia TJ. Reinforcement of a rubbery epoxy polymer by meso structured silica and organo silica with wormhole framework structures. Advanced Functional Materials. 2008;**18**:1067-1074

[52] Gayon E, Debleds O, Nicouleau M, Lamaty F, Van derLee A, Vrancken E, et al. Transition-metal-catalyzed uninterrupted four-step sequence to access trisubstituted isoxazoles. The Journal of Organic Chemistry. 2010;**75**:6050

[53] Fioravanit S, Morreale A, Pellacani L, Tardella PA. Tetrahedron Letters. 2003;**44**:3031

[54] Chakraborty B, Sharma PK, Chhetri MS. Journal of Heterocyclic Chemistry. 2012;**49**(5):1260

[55] De Vitis L, Florio S, Granito C, Ronzini L, Troisi L, Capriati V, et al. Experimental Crystal Structure determination. Tetrahedron. 2004;**60**:1175

[56] Padwa A. In: Katritzky AR, editor. Comprehensive Heterocyclic Chemistry. Vol. III. Amsterdam: Elsevier; 2008. pp. 1-97

[57] Bergmeier SC, Katz SJ, Huang J, McPherson H, Donoghue PJ, Reed DD. Tetrahedron Letters. 2004;**45**:5011

[58] Hodgson DM, Stefane B, Miles TJ, Witherington J. Unsaturated 1,2-amino alcohols and ethers from aziridines and organolithiums. Journal of the Chemical Society, Chemical Communications. 2004:2234

[59] Pourjavadi A, Soleyman R, Bardajee GR. Novel superabsorbent Hydrogel Based on natural hybrid backbone: Optimized synthesis and its swelling behavior. Bulletin of the Korean Chemical Society. 2009;**30**:2680-2686

[60] Danks TN. Microwave assisted synthesis of pyrroles. Tetrahedron Letters. 1999;**40**:3957-3960

[61] Ganji F, Vasheghani-Farahani S, Vasheghani-Farahani E. Theoretical description of hydrogel swelling: A review. Iranian Polymer Journal. 2010;**19**:375-398

Chapter 5

The Effect of Temperature on the Mechanical Performance of Steel and Carbon Fiber Reinforced Polymer (CFRP) Tensegrity System

IfeOlorun Olofin and Ronggui Liu

Abstract

This paper compares the behavioral pattern of steel and carbon fiber reinforced polymer tensegrity system in a suspen-dome that has a span of 4-m span and a 0.4-m, using the finite element method software – namely ANSYS – to undertake the analysis at various temperature regimes. These comparisons were undertaken in order to validate the performance of carbon fiber reinforced polymer cables. Under cold and hot temperatures, the elastic modulus usually reduces as a result of changes in molecular structure. Previous analysis has shown that carbon fiber reinforced polymer cables are able to resist cold and hot temperatures more than steel cables do as the integrity of steel system begins to deform at high temperatures. However, with their low thermal expansion and esthetic properties, carbon fiber reinforced polymer cables can provide structural stability for a tensegrity system in a suspen dome in regions with high temperature conditions.

Keywords: suspen dome, steel tensegrity system, carbon fiber reinforced tensegrity system, temperature effect

1. Introduction

A suspen dome is a long-span roof structure made up of two systems, a reticulated single layer and a tensegrity system created by Kawaguchi and his team. The tensegrity system is made up of three components, namely: the radical cables, struts and hoop cables as illustrated in **Figure 1**. To ensure structural stability, the cables are subjected to tension and the struts to compression. The rigidity of the dome is as a result of the self-stress equilibrium between the cables and the struts.

It is observed in the literature reviewed that a structure with steel material has some setbacks that, hopefully, the use of a new material can overcome, especially since the need for long-span structures is currently trendy. This paper describes the elastic plastic state in respect of temperature changes of carbon fiber reinforced polymer cable as a tensegrity system in a suspen dome in comparison with that of steel cables.

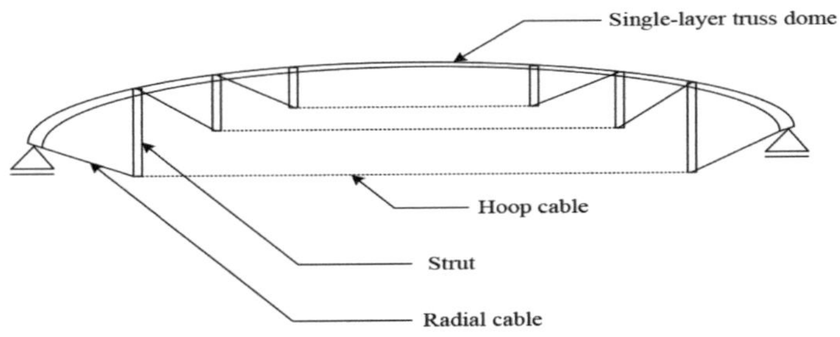

Figure 1.
Suspen-dome (structural system).

2. Literature review

Civil Engineers are seeking new types of materials to create slender structures with outstanding strength in their structural designs for long-span structures. Carbon fiber reinforced polymers are currently being used in various engineering fields based on their high stiffness, high strength and low density. Researchers have investigated that carbon fiber reinforced polymer cables can be used to strengthen steel members both with numerical and experimental methods [1, 2]. However, based on literature review, it is clear that researchers have not provided guidelines for its implementation in space structures such as suspen dome. Early researches focused on static and dynamic effect of fire, and joint treatment [3–13] of suspen dome structures with steel materials for numerical and experimental methods. Of course, steel materials can effectively enhance the strength of structures, but a larger amount of strengthening materials would be required to improve serviceability of such structures.

Research on the environmental durability of carbon fiber reinforced polymer compared to that of steel bond is limited, particularly in civil engineering and infrastructure disciplines [1]. The relatively high modulus of carbon fiber reinforced polymer materials can enhance their serviceability and ultimate better load-bearing capacity than those of steel structures [1]. Also, since the effect of temperature rise is included in structural models which describe the structural behavior, stability of elements and material properties, the incremental reduction in steel strength with increase in temperature, a well-known phenomenon, creates the room for replacement with more resistant materials such as carbon fiber reinforced polymer. This is especially so because structures exposed to sunshine react to temperature changes. Thus, temperature exposure is one important scenario to consider when selecting construction materials. As long as the rising temperature remains below the debonding temperature threshold, the reinforcing material will retain its function.

Al-Salloum et al. [3] investigated the load-bearing capacity of carbon fiber reinforced polymer reinforced specimen exposed at 100°C, using a cylinder $\phi 100 \times 200$ mm strengthen with a single layer of carbon fiber reinforced polymer, and the results showed that, at the specified temperature, the load-bearing capacity of the carbon fiber reinforced polymer reduces slightly. Notwithstanding this, with the continued development and application of carbon fiber reinforced polymer cables, researchers are becoming more confident in its use for bridge engineering. The brittle nature of carbon fiber reinforced polymer cables could be seen as a disadvantage for its application for a long span structure such as the tensegrity system of a suspen-dome; however, further researches in other aspects, such as temperature changes, are required to determine its effectiveness and advantage over steel cables. Steel is brittle with temperature ranging between −72 and −40°C and its ductile-brittle transition ranges from −40 to 0°C. Carbon fiber reinforced polymer

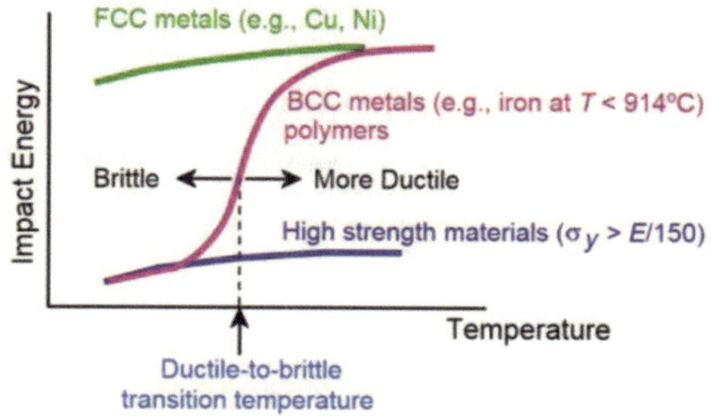

Figure 2.
Ductile-to-brittle relation.

is of high strength and brittle so it does not display a ductile to brittle transition. Steel exhibits a ductile to brittle transition and can only be applied at temperatures above the ductile-brittle (**Figure 2**) transition threshold to avoid failure without warning. Carbon fiber reinforced polymers rarely have a ductile-brittle transition threshold; hence it can be used at any temperature while steel can be used only at temperatures above 0°C.

According to Zhang et al. [14], temperature effect has a great influence on the mechanical properties of a suspen dome roof. Also, the results of Chen et al. [10] further indicated that different thermal expansion coefficients of materials will significantly affect the mechanical properties of cables that form a part of a suspen dome roof structure. Since the temperature expansion coefficient of carbon fiber is about 1/18 of that of steel cables, the influence of temperature effect of carbon fiber reinforced polymer obviously requires further investigation to determine if its temperature effect will have satisfactory impact on a suspen dome structure.

2.1 Theoretical formula on pre-stress losses based on temperature effect

Evaluation process of pre-stress loss affected by temperature is given as:

$$\Delta L = \alpha L\, o(t - t_o) \tag{1}$$

$$\varepsilon = \Delta L/L = \alpha(t - t_o) \tag{2}$$

$$\varepsilon_x = \varepsilon_{xx} + \varepsilon_{xy} = \alpha_L(t - t_o) + \nu_{\min}\alpha_T(t - t_o) = (\alpha_L + \nu_{\min}\alpha_T)(t - t_o) \tag{3}$$

$$\varepsilon_{xy} = \nu_{yx}\varepsilon_y \tag{4}$$

$$\sigma_x = \varepsilon_x E \tag{5}$$

$$\Delta N = A\,\sigma_x \tag{6}$$

where: ΔL – deformation length of cable, L – the initial length of the cable, α_L – Longitudinal linear expansion coefficient, α_T – transverse linear expansion coefficient, ε_x – the total strain in the x-direction, ε_{xx} – the strain component in the x-direction due to deformation in the x-direction, ε_{xy} – the strain component in the x-direction due to deformation in the y-direction, $\nu_{\min}$ – times Poisson's ratio, ν_{yx} – Poisson's ratio in the y-direction of the material, ε_y – the total strain of the cable in the y-direction, E – effective elastic modulus of cable, A – cross-sectional area of the cables.

3. Description of proposed model

The model in the study is shown in **Figure 3a** and **b**. It has a span and rise of 4 and 0.4 m, respectively (making a rise-to-span ratio of 0.1 that satisfies the span to rise ratio recommend by Kitipornchai et al. [4]).

Basic assumption

- The tensegrity members are pin-jointed.
- The external loads are applied at nodes.
- The self-weight is transferred to nodes as point loads.
- Cables are elastic

The single-layer (upper section of the suspen-dome structure) consists primarily of hollow beams which are 20 mm in diameter and 1.2 mm in thickness. A large cross-section gives a high stiffness which contributes to the overall stability of the shells. In this study, the uniformly distributed load is converted to nodal loads at all nodes on top of the suspen dome. For example, a uniformly distributed load 3 kN/m^2 is converted to nodal load of magnitude 20 kN. A distributed load of 20 kg was assumed to be the dead load applied on the roof, including the loads on the cables. The initial force was designed to increase as external loads increase to avoid slacking when the structure is in service. If the loads are increased beyond a certain limit, the structure will collapse suddenly under any little disturbing force. Hence, any designer of a new structure should be interested in knowing the load factor value which would cause the structure to collapse. Subramanian [5] stated that the detailed information about the behavior of the structure at or near the critical load is insignificant for practical purposes. As a result of this statement, different loading conditions were not considered in the investigation. For numerical reasons the temperature loads were taken to be −30 and 30°C (cold and hot levels). Any valve of ΔT can be applied because the assumption of linearity is considered.

A typical numbering of nodes and elements illustrated in **Figure 4** due to the symmetrical nature of the structure where E*i* represents element and *i* number range from 1 to 5 and N*i* represents node, *i* a number ranging from 1 to 5, respectively.

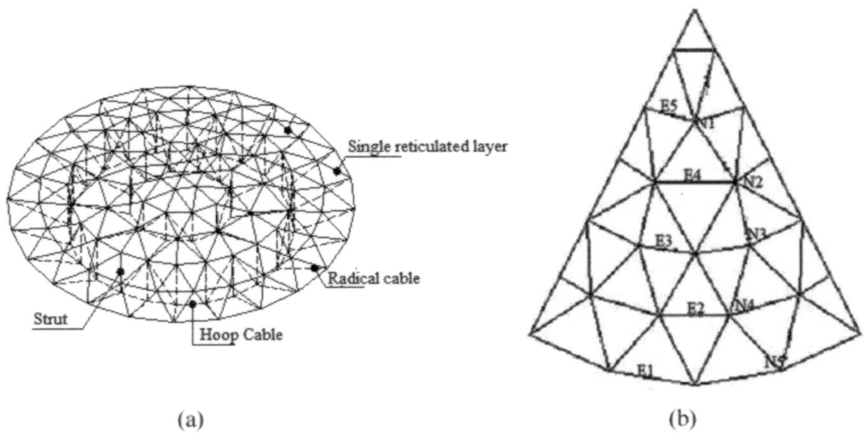

Figure 3.
Details of the model. (a) A sketch of the geometry of the suspen dome. (b) Numbering of nodes and element.

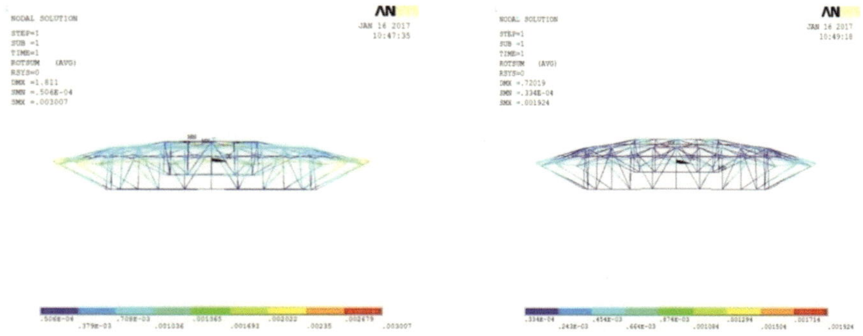

Steel(-30°C)

Carbon fibre reinforced polymer (-30°C)

Figure 4.
Comparison of steel and CFRP tensegrity at −30°C.

	Steel cables	Carbon fiber reinforced polymer cables	Steel section
Modulus of elasticity(N/mm^2)	1.8×10^5	1.6×10^5	2.05×10^5
Poisson's ratio	0.3	0.3	0.3
Thermal expansion (K^{-1})	1.2×10^{-5}	0.68×10^{-7}	1.2×10^{-5}
Tensile strength (MPa)	835	2550	

Table 1.
Properties of the materials.

3.1 Description of the properties of materials

In this section the values of the parameters of the materials and those of the materials used for each element of the system are described. The properties of the materials for the cables and single reticulated layer dome are illustrated in **Table 1**.

4. Results and discussions

The use of tensegrity system provides an interesting solution to the stability of a suspen dome. This has prompted the proposal of using carbon fiber reinforced polymer cables. The results are based on analogy and can satisfactorily anticipate the behavior of the structure.

4.1 Effect on structural member based on numerical study

For steel structure at both cold and hot temperature, shown in **Figures 4** and **5**, the member bars of the upper single layer reticulated shell at mid rib stress increase due to the pulling force of radical cables, strut and hoops cables. It must be noted that when the cables are loose at the temperature load, the suspen dome cannot maintain its structural form and a collapse takes place due to some cables being slack and the strut being in tension.

At the outermost ring, the displacement is smaller due to the constraint boundary condition which minimizes stress in its member and the further the nodes are away to the mid rib the larger the displacement. The inner cables are of relatively large deformation. Pre-stressed cable has an impact on the behavior of a suspen

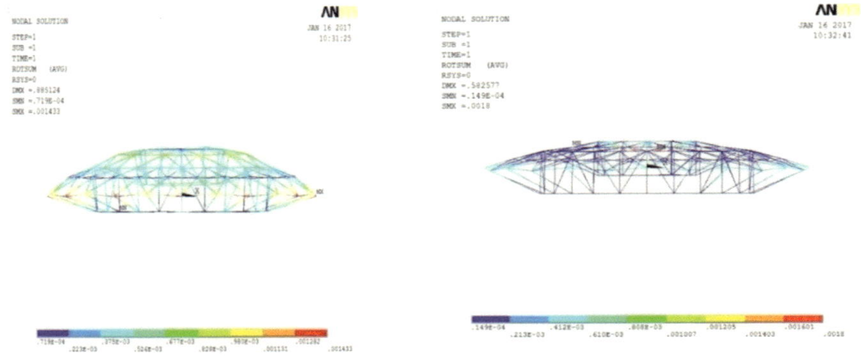

Steel(+30°C)

carbon fibre reinforced polymer (+30°C)

Figure 5.
Comparison of steel and carbon fiber reinforced polymer tensegrity at +30°.

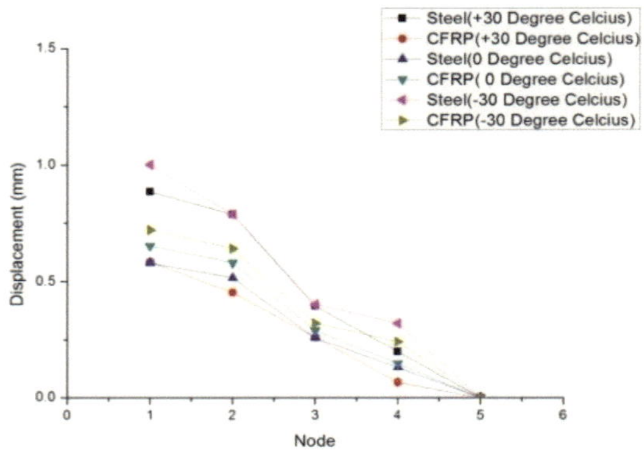

Figure 6.
Displacement against nodes.

dome structure, however with the increase in temperature of steel cable, the stiffness and load bearing capacity reduced drastically.

For the carbon fiber reinforced polymer tensegrity system at both cold and hot temperature, the cables tend to be in tension resisting the pushing force from the single layer reticulated shell to avoid slack in cables and tension in struts, as shown in **Figures 4** and **5**. Deformation of the structure in both cases is due to stress, a tensile or compressive stress that is a result of the thermal expansion.

For steel cables, the stress-strain effect increases tremendously with increase in temperature value which causes the structure to expand thereby reducing its serviceability. However, the difference in displacement effect for carbon fiber reinforced polymer cable was minute as shown in **Figure 6** which implies that a steel material used as the single reticulated layer expands and deforms with cold and hot temperature changes whereas carbon fiber reinforced polymer cables are constrained to avoid deformation.

4.2 Effect on single reticulated layer

The stresses of the members are not zero as shown in **Figure 7** which indicated the structure has not reached its bearing capacity threshold under the given load. The results also demonstrate that at hot and cold temperatures, carbon fiber

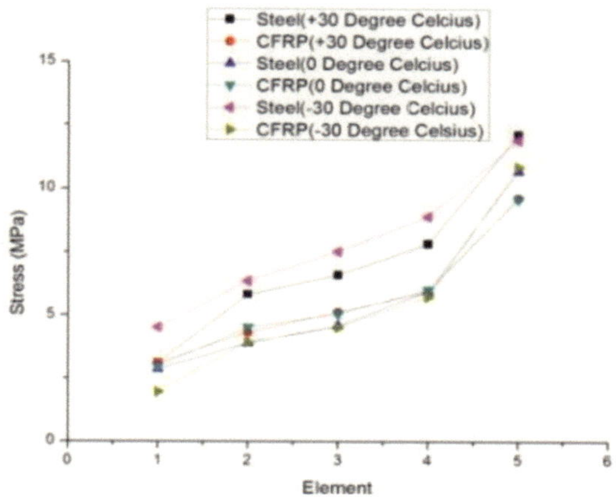

Figure 7.
Axial load vs. element number curve.

Cable	Steel	Carbon fiber reinforced polymer
Hoop cable	333.15	310.57
Radical cable	352.10	308.06

Table 2.
Internal force in tension members at +30°C (N).

Cable	Steel	Carbon fiber reinforced polymer
Hoop cable	365.45	304.57
Radical cable	353.42	301.06

Table 3.
Internal force in tension members at −30°C (N).

reinforced polymer tensegrity system maintains the same minimal stresses in members compared to steel tensegrity system.

4.3 Effect on pre-stress losses on the cable members

Tables 2 and **3** illustrate maximum internal forces generated in the outer hoop and radical cable of the structural system due to temperature effect. The internal forces for both the hoop and radical carbon fiber reinforced polymer cable are minimal compared with steel.

Hence, an overall stability with reference to temperature is guaranteed for a suspen dome with carbon fiber reinforced polymer tensegrity system.

5. Feasibility consideration

The numerical results of this study show that the proposed application of carbon fiber reinforced polymer cables for a tensegrity system in a suspen dome is more effective compared to steel tensegrity with respect to temperature effect. The results also show that a composite material has less adverse effect on the stress in the structural members in terms of temperature changes. Fureai dome has been constructed using the suspen dome system in Nagaro prefecture, where there is heavy snow fall. It is proposed to measure the changes in cable tension force due to the snow fall and temperature changes [5]. With the results obtained, it is suitable for situations where the temperature factor is considered. Thus the traditional suspen dome roof structure of steel cables can be replaced with carbon fiber reinforced polymer cables because they can achieve better design results than those of steel.

6. Conclusion

This study has shown that carbon fiber reinforced polymer cables have displayed desirable traits which can be exploited in the design of a tensegrity system. The deformation attribute in terms of temperature variation is small for the carbon fiber reinforced polymer tensegrity system hence changes in geometry can be neglected compared to steel with high displacement values under different temperature conditions.

Although there has been no relatively practical application of the concept, the results obtained in this study indicate that carbon fiber reinforced polymer cables can be recommended for use because of for its desirable effects. The objective of this study, using a small suspen dome model to investigate the influence of temperature changes on the stability of a suspen dome, has been achieved satisfactorily. Thus, this paper demonstrates that the proposed carbon fiber reinforced polymer system can be effectively used to enhance the serviceability and stability of a suspen dome.

Acknowledgements

The authors acknowledge the support of the National Science Foundation of China (Grant no: 51608234) and Province Science of Jiangsu: BK20160534.

Conflict of interest

The authors declare no conflict of interest.

Author details

IfeOlorun Olofin* and Ronggui Liu
Department of Civil Engineering and Mechanics, Jiangsu University, Zhenjiang, Jiangsu, P.R. China

*Address all correspondence to: epher2002@yahoo.com

References

[1] Dawood M, Rizkalla S. Development of a Carbon Fiber Reinforced Polymer System for Strengthening and Repair of Steel Bridges and Structures, 2nd Canadian Conference on Effective Design of Structures. Ontario, Canada: McMaster University Hamilton; 2008

[2] Xu X, Xiao Zhang L, Yonggang S. Static and dynamics characteristic of a long span cable-stayed bridge with CFRP cables. Materials. 2015;7:4854-4877. DOI: 10.3390/ma7064854

[3] Al-Salloum YA, Elsanadedy HM, Abadel AA. Behavior of FRP confined concrete after high temperature exposure. Construction and Building Materials. 2015;**25**(25):835-850

[4] Kitipornchai S, Kang WJ, Lam H, Albermani F. Factors affecting the design and construction of lamella suspen dome system. Journal of Constructional Steel Research. 2005;**61**:764-785

[5] Subramanian N. Space Structures: Principles and Practice. Brentwood, Essex, UK: Multi-Science Publishing Co Ltd; 2006

[6] Wenjiang K, Zhihua C, Heung-fai L, Chenian Z. Analysis and design of the general and outmost ring stiffened suspen dome structure. Engineering Structures. 2003;**25**:1685-1695

[7] Zong Z, He Y, Li S. Analysis of the fire resistance of a suspen dome structure. Applied Mechanics and Materials. 2012;(**166-169**):1040-1044

[8] Zhang ZH, Cao QS, Fu XY, Dong SL. Structural design of the large span suspen dome. In: Computational Mechanics. 2009. p. 270

[9] Chen ZH, Sun GJ, Liu ZS. The effect of cable material characteristic on suspen dome. Advanced Materials Research. 2010;**156-157**:1251-1255

[10] Chen ZH, Wu YJ, Yin Y, Shen C. Formulation and application of multi-node sliding cable element for the analysis of suspen-dome structures. Finite Elements in Analysis and Design. 2010;**46**: 743-750

[11] Chen Z, Yan R, Wang X, Liu H, Xiao X. Experimental researches on suspen dome structure with rolling cable-strut joint. Advanced Steel Construction. 2015;**11**(1):15-23

[12] Zhi-qiang L, Zhi-hong Z, Shi-lin D, Xue-yi F. Construction sequence simulation of a practical suspen-dome in Jinan Olympic Centre. Advanced Steel Construction. 2012;**8**:38-53

[13] Mamoru K, Masaru A, Tatsuo H, Ikuo L, Satoshi F, Hiroaki M, et al. Structural tests on the system dome system. ASCE Journal. 1994:383-392

[14] Zhang A, Liu X, Cui W, Rao W. Analysis of the overall stability of suspen dome structure with temperature change. Steel Construction. 2009. (In Chinese)

Chapter 6

Thick-Section Epoxy Composites

Yanan Hou, Liguo Li and Joseph H. Koo

Abstract

Thick-section composites (TSC) are extensively demanded in many fields, such as aerospace, wind energy, and oil and gas industries. However, the manufacturing process of thick-section thermoset composites (TSSC) encounters significant complexities, such as variations of nonuniform resin flow, exothermal reaction and curing, and dimensional stability through the thickness direction. These process-related nonuniformities are expected to result in through-thickness gradients of mechanical properties and curing-induced deformations, leading to undesirable residual stresses and damage. This chapter introduces the application of TSC and issues related to its manufacturing processes. Methods of TSC are examined and analyzed. Fundamental characteristics of curing kinetics, thermal transfer, and residual stress in TSC will be explained. Research of detailed experiments will be referred for readers for further studies.

Keywords: thick-section, thermoset, composite, glass fiber, carbon fiber, crack, micro-crack, manufacturing, residual stress, coefficient of thermal expansion, heat transfer, wrinkles, porosity, wind energy, wind turbine blade, oil and gas, downhole tools, curing kinetics, curing shrinkage, waste disposal, DSC, DEA

1. Introduction

To differentiate thin-section composites and composite film, fiber-reinforced composites (FRC) with cross-sections of 1 inch or thicker are generally known as TSC, regardless of the geometry of the part. Well-known of its low cost and high strength-to-weight ratio, TSC are widely used in aerospace, wind energy, oil and gas industries, and others. For examples, a 100-meter-long offshore wind turbine blade can be made of FRC with a cross-section as thick as 3 inches. A composite oil tool can require a thickness up to 2 inches. However, the manufacturing process of TSSC encounters significant complexities, such as variations of nonuniform resin flow, exothermal reaction and curing, and dimensional stability through the thickness direction. These process-related nonuniformities are expected to result in through-thickness gradients of mechanical properties and curing-induced deformations, leading to undesirable residual stresses and damage. This chapter introduces the application of TSC and issues related to its manufacturing processes. Methods of TSC will be examined and analyzed. Knowledge on the fundamental characteristics of curing kinetics, thermal transfer, and residual stress in TSC will be explained. This chapter focuses on method explanation. Researches of detailed experiments will be referred for readers to further studies.

2. Application of TSC

Due to its high strength-to-weight ratio, composite material is widely applied in aerospace industry, wind energy, oil and gas, automotive industry, sports, and construction and agricultural equipment. Some of the state-of-the-art applications include high-temperature composite material that developed to work in extreme environments up to 700°F continuously could be potentially used by engine original equipment manufacturers (OEMs). Carbon fiber/epoxy rebar is developed to replace steel for construction. Sprayer boom arms made of carbon fiber/epoxy could range from 100 to 150 ft. long for agricultural equipment. In aerospace industry, aircraft interiors are widely made of epoxy pre-preg material. In the world's largest commercial aircraft, A380, 16% of the structural weight is made of advanced polymer composites, where the carbon fiber composite center wing box weighs over 9 tons. Fuselage panels and stabilizers add another 4% composite by weight [1]. In oil and gas industry, composite, other than metal, is increasingly used in pipeline to lower its weight and increase its flexibility. Composite downhole tools are favored due to their ease of disposal. In sports, composite material is found in bike frames, kayak paddles, dog handles, and more. In wind energy, fiber-reinforced composites are utilized in wind turbine rotor blades. The traditional automotive industry is inertia to replace metal by composites, but the emerging electric vehicle market offers opportunities to composite manufacturing technologies, including carbon fiber/epoxy sheet molding compound (SMC) wheels, lift gates, and composite battery enclosures. Filament wound liquefied petroleum gas (LPG) tanks are used for industrial and home gas storage. Pultruded composites are used as ballistic panels to protect school.

Besides replacing metal and wood, composite could also be a potential substitute of natural rocks due to its long-lasting nature. For example, stone carving is believed to be the most reliable method to storage information, which is able to retain its data for thousands of years, compared with paper that could last for hundreds of years or hard drive for less than 20 years in well-maintained storage environments. Polymer composites, coupled with laser or water printing technologies, might be used as a media to record information and be stored under water, for example, in the sea. Researchers at Virginia Tech's College of Science are also working on porous carbon fiber made from block copolymers, similar to a sponge, to store ions of energy [2].

As the world population reached 7.6 billion and expected to almost double in the next hundred years, many composite equipment are required to become larger and thicker. For example, the global offshore wind energy market grew by 0.5% in 2019, with new installations of 4.49 gigawatts [3]. A common seen wind turbine in the USA is Gamesa G87 made in Spain with turbine blades of 43.5 meter long. Its capacity is 2 megawatts. Wind turbines of increasingly larger capacities are demanded. A prototype 100-meter-long offshore wind turbine blade made of FRC has a cross-section up to 3 inches thick. Another example is in aerospace; Northrop Grumman is adapting its automated stiffener forming technology to build spars. Thickness of most stiffeners is less than 0.25 inch, but that of spars can be 1 inch or thicker [4], which will dramatically increase the complexity of the manufacturing process.

3. Thick-section epoxy composite (TSEC)-related issues

3.1 Micro-cracks

Comparing to thin-section composites, the manufacturing of TSEC introduces superimposed complexities. Micro-cracks are often found in TSC, which largely

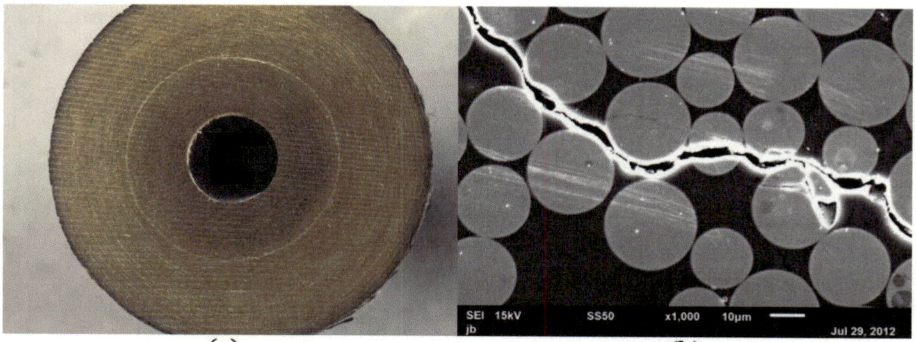

(a) (b)

Figure 1.
TSC with micro-cracks. (a) TSC with micro-cracks. (b) Micro-cracks under SEM.

deteriorate the consistency of its mechanical properties. Micro-cracks can be visually examined as shown in **Figure 1(a)**. A graph of micro-crack under scanning electron microscopy (SEM) is shown in **Figure 1(b)**.

Micro-cracks are mainly caused by the internal stress generated during the curing of the composites and the cooling process. The internal stress of TSEC is affected by the following parameters.

3.1.1 Curing shrinkage

Epoxy is a thermoset polymer generally superior to polyester and vinyl ester in terms of mechanical properties and cost. Epoxy resin can be thermally cured or cured with a wide range of hardeners at room temperature or elevated temperatures, including amine, anhydrides, phenols, alcohols, and thiols. Cross-link reactions of epoxy accompany with curing shrinkages. The shrinking is a continuous process along with the increase of the degree of curing of the material. For example, a standard bisphenol A epoxy (e.g., Epon 828) shrinks 2.8% pre-gelation and 2.5% post-gelation [5]. Curing shrinkage of epoxy can be up to 9%.

3.1.2 Nonuniformity of heat distribution

Curing speed of epoxy is positively related to temperature. Meanwhile, the cross-linking reaction releases heat, which adds an additional item to the already complex heat transfer equation. The main author's earlier research [6] shows the temperature difference of a 3.6-inch-thick FRC can be up to 52°C in the thickness direction, with the hottest layers in the middle of the part. The temperature gradient leads to a curing degree difference of up to 70% in the thickness direction. This implies when the epoxy resin in the middle of the part is solidified, epoxy resin at outer surfaces is still in liquid state. While the epoxy resin at the outer surfaces is solidified, the accompanying volume shrinkage generates stress among the surface layers and the already solidified middle layers. The residual stress can be high enough to create micro-cracks among layers. Complexities piles up when dealing with epoxy systems requiring elevated temperature curing cycle. For example, a filament wound TSC billet may take 4 hours to wind, resin in layers closer to inside diameter (ID) cures hours earlier than the outer layers. Once the billet is moved into an oven after winding for further curing, it may take a couple of hours for the heat transferring to the core. Outer layers cure faster than the inner layers at the beginning, and the middle layers come from behind later due to heat overshoot caused by the reaction. As a result, micro-cracks often occur between middle layers and the ID or between middle layers and the surface, as shown in **Figure 1**.

3.1.3 Coefficient of thermal expansion (CTE)

Thermal expansion coefficient of epoxy is over 10 times higher than that of glass or carbon fibers that are widely applied as reinforcement in FRC. A curing profile of FRC usually includes several steps of heat soaking and cooling down, during which, bonds between the resin and its reinforcement face repeated stress and fatigue, where failure leads to interlaminar cracks. Tooling complexes the problem with an additional thermal expansion coefficient in resin transfer molding (RTM) or filament winding (FW) processes.

3.2 Mechanical property nonuniformity

Unlike most metallic materials with isotropic properties, FRCs often exhibit anisotropy properties. For example, a glass fiber-reinforced TSC laminate displaces various mechanical properties in different directions. As shown in **Figure 2**, tensile strength in the fiber direction (LZ direction) is over 30 times higher than that in the thickness direction (ZT direction) [7].

Tensile strength of glass fibers (e.g., 321–343 Ksi for a commercial E glass fiber) is over 34 times higher than that of the matrix epoxy (e.g., 9.7 Ksi for a bisphenol A epoxy), and the Young's modulus of glass fibers (12.1×10^6 psi for a commercial E glass fiber) is over 25 times higher than that of the matrix epoxy (0.45×10^6 psi for a cycloaliphatic epoxy cured with an anhydride curing agent). Mechanical properties of FRC along fiber directions are dominated by properties of fibers; those that are transverse to fibers are weakened by resin matrix and bonds between epoxy and fibers.

Mechanical properties of epoxy are determined by its cross-linking density, which is controlled by its curing profile. For example, a rapid cure may result in a quick but sloppy cross-linking, and a low curing temperature may leave the epoxy not fully cured. In theory, a slow curing process with a rationally higher maximum curing temperature is always preferred, but it can be costly and infeasible in a production scale.

Glass transition temperatures (T_g) are often used to imply mechanical properties of post-gelled epoxy. T_g represents the temperature where epoxy transits the fastest

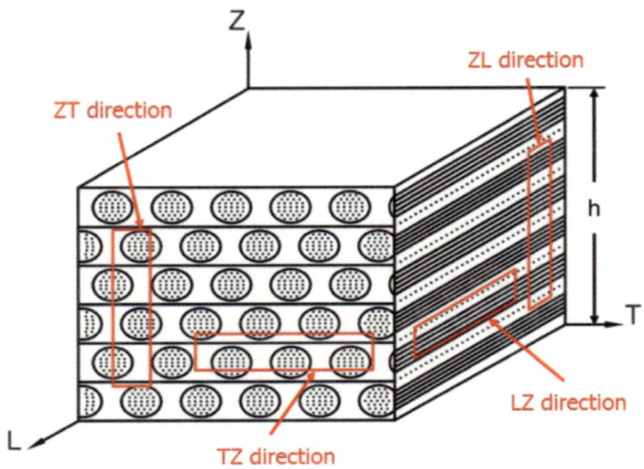

Figure 2.
Schematic diagram of continuous fiber-reinforced TSC.

from "glassy" state to rubbery state. The main author's earlier research [6] shows the T_g difference of a 3.6 inch TSC laminate can be up to 15°F. T_g differences are caused by temperature gradients in the TSC part during the curing process, and the variation of mechanical properties lowers the quality of the composite part.

3.3 Wrinkles and porosities

Epoxy, with trapped air, flows among fibers before gelation, causing relaxation of fibers and dry spots. This often results in wrinkles and porosities, especially in TSCs. Wrinkles and porosities dramatically deteriorate matrix-dominated properties and can severely limit the performance of the TSC parts. Researches show lower curing pressure reduces wrinkles but may increase porosities [8]. Correct design of process parameters and accurate measurements are essential to quantify and eliminate manufacturing defects. For example, in FW, a gap between feeding fiber bands creates fiber waviness and results in pineapple-like skin, as shown in **Figure 3**.

Porosity is a major concern of TSC manufacturing, especially of pre-preg/tow-preg processed TSC. Fibers pre-coated with resin are laid up to required thickness and shape before curing. During the curing process, coating resin is liquefied, filled in gaps between fibers, and then solidified. Trapped air among fibers requires time to escape before the gelation of resin. Thus, TSC made with pre-preg/tow-preg process often yields relatively higher porosity. **Figure 4** shows the void under SEM.

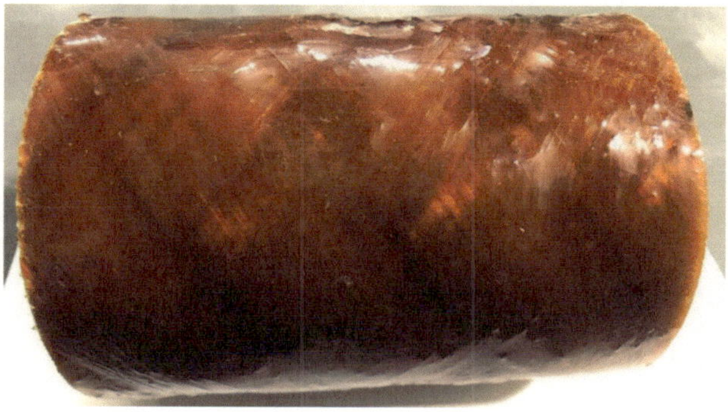

Figure 3.
TSC with pineapple-like skin.

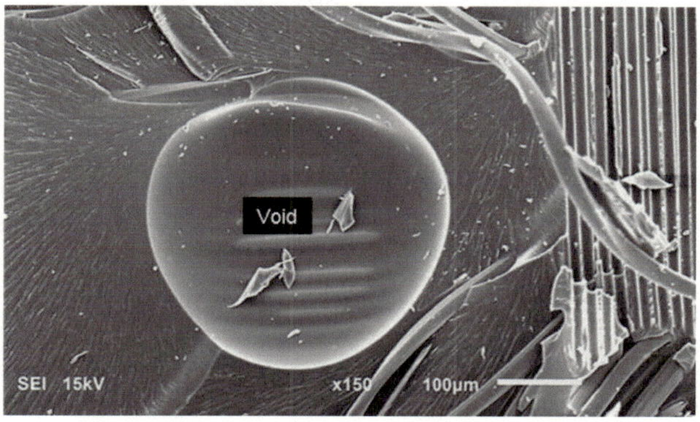

Figure 4.
Void in TSC under SEM.

3.4 Waste disposal and heat overshoot

Gelation happens after either mixing resin and hardener or adding catalyst into homopolymerizing resin. Excess mixed epoxy or epoxy squeezed out during processing is hard to recycle and turns out to be waste. Even with an effective waste reducing plan, the quantity of production epoxy waste from TSC manufacturing is often considerable. Disposing the waste under the Resource Conservation and Recovery Act (RCRA) rules can be costly and complicated. The waste can be categorized into three scenarios [9]:

- Disposal resin, hardener, or catalyst before mixing. It's often ignitable hazardous waste (EPA waste code D001) or corrosive waste (EPA waste code D002). A waste company specialized in chemical disposal is required to handle the waste, which can cost around $25 per gallon, plus transportation and fuel.

- Disposal mixed resin that is not fully cured. It may be subject to both solid waste (40 CFR 261.3) and unused commercial chemical product (40 CFR 261.33). It may be on the P and U lists of chemical product waste, and none of the F or K lists for specific and non-specific sources of hazardous waste will apply.

- Disposal fully cured epoxy. This scenario is the easiest to deal with. Fully cured epoxy rarely exhibits hazardous waste characteristic and can be disposed as general solid waste.

In the industry, pure resin/hardener or unfully cured waste is usually stoichiometrically mixed with hardener/resin and fully cured before disposal. A waste disposal curing profile should be carefully designed to avoid reaction overheat and burning accidents.

4. Methodologies of TSEC analysis

4.1 Curing kinetics characterization

Differential scanning calorimeter (DSC) is often used to characterize the kinetics of thermoset resins. Data can be used to study the relationship between curing degrees and temperature. Kamal-Sourour relationship is usually used in the curing kinetics study of pure bisphenol A epoxy, as shown in Eq. (1).

$$\frac{da}{dt} = ka^m(1-a)^n \tag{1}$$

where a is the degree of cure, T is the instantaneous temperature, m and n are orders of the reaction, and k is the rate of constant, following the Arrhenius relationship.

$$k = Aexp\left(-\frac{\Delta E}{RT}\right) \tag{2}$$

The ΔE in Eq. (2) is the activation energy of the epoxy, A is the frequency factor, and R is the universal gas constant. To obtain the material parameters, three DSC isothermal tests are usually conducted. The three temperatures can be selected in the following way:

1. Conduct a temperature scanning on the epoxy of interest at 5°C/min using DSC.

2. Select the reaction initiation temperature as the first temperature, which is implied by the start of an uptrend of the heat flux curve. Isothermal test at this temperature captures cure kinetics characterization of the beginning of the curing reaction.

3. Select the second temperature by drawing tangential lines, one along the heat flux curve before reaction starts and one at the fastest reacting point before the first peak.

4. Select a temperature a little lower than the first peak temperature as the last temperature. Isothermal test at this temperature starts fast. It tends to lose track on some heat at the beginning of the reaction, but captures characterization of the cure kinetics at the latter part of the reaction.

For example, **Figure 5** shows heat flux of a commercial resin from a DSC temperature scanning at 5°C/min. The three temperatures for isothermal tests can be 22, 50, and 70°C.

Degree of cure and curing rate can be calculated from DSC data, and three curves of curing rate versus degree of cure can be plotted at each test temperature (T1, T2, and T3). The reaction orders and the rate of constant in the reaction kinetics model can be determined by a nonlinear fitting method using either curing kinetics software or MATLAB. The activation energy (ΔE) and lnA can be obtained by a linear fitting of lnk and $1/T$.

Bailleul's [10] equation is often used for more complex polymerization reactions, as shown in Eq. (3).

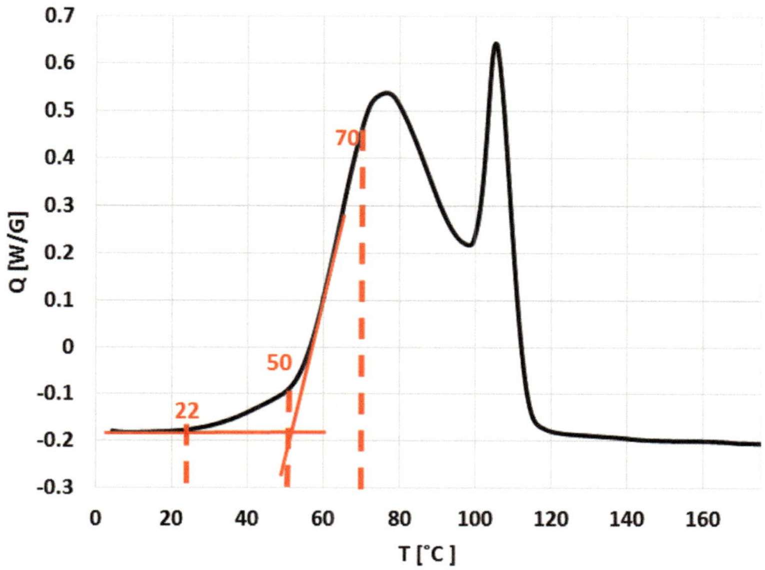

Figure 5.
DSC experiment for isothermal temperature selection.

$$\frac{da}{dt} = W(v(T)) \cdot K(T) \cdot G(a) \cdot F_{diffusion}(a) \tag{3}$$

where $W(\nu(T))$ represents the induction of reaction or the inhibition period when an inhibitor is added to the resin. It defines as below:

$$W\left(v(T)\right) = \begin{cases} 0, if\ v(T) \geq 0 \\ 1, if\ v(T) < 0 \end{cases} \tag{4}$$

$v(T)$ is expressed as Eq. (5).

$$v(T) = t_{ref} - \int_0^{t_{ind}} exp\left[-B \cdot \left(\frac{T_{ref}}{T} - 1\right)\right] dt \tag{5}$$

$K(T)$ represents the dependence of curing rate on temperature, as shown in Eq. (6), $G(a)$ represents the dependence of curing rate on the degree of cure as shown in Eq. (7), and $F_{diffusion}(a)$ describes the diffusion control in the curing, as shown in Eq. (8).

$$K(T) = K_{ref}\, exp\left[-A\left(\frac{T_{ref}}{T} - 1\right)\right] \tag{6}$$

$$G(a) = \sum_{i=0}^{6} G_i a^i \tag{7}$$

$$F_{diffusion}(a) = \frac{1}{1 + exp\left(E_1(a - a_c)\right)} \tag{8}$$

The equation fits well in the reaction of DERAKANE MOMENTUM 411–350 epoxy vinyl ester and 1 wt% NOROX MEKP-925H (MEKP stands for methylethylketone peroxide) with 0.05 wt% of cobalt naphthenate well [11].

In situ degree of curing can be quantified by dielectric cure analyzer (DEA). DEA sensors can be embedded during the manufacturing process of TSC and measures in situ ion viscosities (ρ) of the matrix resin. Resin ion viscosity is positively correlated with its degree of cure and negatively correlated with temperature. Curing index, which is equivalent to degree of curing, is calculated as Eq. (9).

$$CI = \frac{log\left[\sigma_{f_0}(a, T_0)\right] - log\left[\sigma_{f_0}(0, T_0)\right]}{log\left[\sigma_{f_0}(1, T_0)\right] - log\left[\sigma_{f_0}(0, T_0)\right]} \times 100\% \tag{9}$$

where σ is the ion conductivity of the material, which is the reverse of ion viscosity. The subscript f represents the frequency of the voltage of excitation used in the DEA test. The temperature and frequency with the subscript 0 refer to the testing frequency and temperature of the curing index calculation. The main author's previous study [12] showed well agreement of curing index and degree of cure in AOC Altek R920-E polyester mixed with a 2% NOROX MEKP-925 catalyst and glass fiber-reinforced composites.

In TSC analysis, thermocouples are usually placed at locations of interest to record the temperature history during manufacturing processes, including pre-curing process, curing process, and post-curing process. Curing degree profiles at various locations are calculated by inputting temperature history into the predetermined curing kinetics model. In situ curing index profiles can be measured using DEA sensor and related to degrees of cure. The degree of cure profiles can be used to optimize the curing profile of the composites and evaluate properties of the end composite part. For example, the glass transition temperature can be modeled through a function of degree of cure developed by Pascault et al. [13], as shown in Eq. (10).

$$\frac{T_g - T_{g0}}{T_{g\infty} - T_{g0}} = \frac{a \cdot k}{1 - (1 - k) \cdot a} \tag{10}$$

where T_{g0} is the glass transition temperature of raw epoxy and $T_{g\infty}$ is the glass transition temperature of fully cured epoxy. k is a structure-dependent parameter that can be obtained by nonlinear fitting of T_g and curing degree data from DSC tests.

Curing degree profiles also provide essential parameters for further curing shrinkage and inner stress study.

4.2 Heat transfer characterization

Heat transfer of thermoset composite differentiates itself from metal materials primarily on three aspects: (1) changing specific heat capacity, (2) changing thermal conductivity coefficient, and (3) reaction heat.

4.2.1 Specific heat capacity

Specific heat capacities of composite are usually calculated using mixing law as shown in Eq. (11).

$$C_{p-composite} = C_{p-fiber} \cdot w_f\% + C_{p-epoxy} \cdot \left(1 - w_f\%\right) \tag{11}$$

where w_f % represents the weight percentage of reinforced fiber.

Specific heat capacities of epoxy depend on temperature and degree of cure, where a mix law may also be established, as shown in Eq. (12).

$$C_{p-epoxy}(T, a) = C_{p0}(T) \cdot a + C_{p1}(T) \cdot (1 - a) \tag{12}$$

where C_{p0} is the specific heat capacity of raw epoxy, C_{p1} is the specific heat capacity of fully cured resin, and a is the degree of cure.

Modulated DSC tests are usually used to capture reversible heat flux of thermoset resin, which can be used to calculate the specific heat, as shown in Eq. (13).

$$C_p(T) = \frac{-q}{m \cdot dT/dt} \tag{13}$$

where q represents heat absorbed by the material to increase one-unit degree of temperature, m represents the mass of the material, and T represents the temperature. For unreacted epoxy, DSC scanning includes several plateaus at temperatures below the reaction initiating temperature. Three heating speeds are normally applied. 5°C/min, 10°C/min, and 20°C/min are normally selected. A curve fitting of specific heat capacity versus temperature yields the relationship between specific heat capacity of unreacted resin and temperature. The relationship between specific heat capacity of fully cured resin and temperature can be obtained through similar DSC scanning with plateaus including several temperatures higher than the glass transition temperature of the material and several temperatures lower. For example, Bailleul et al. [14] selected 30, 70, and 100°C for the raw material and 30, 70, 110, 150, and 190°C for the fully cured resin, where the initiation temperature of the resin is higher than 100°C and the T_g of the fully cured resin is about 130°C. In Bailleul et al.'s research, curve fittings present linear relations between the heat specific capacity and temperature for all three scenarios: raw resin, cured resin before the glass transition period, and cured resin after the glass transition period. For fully cured resin during the glass transition, the specific capacity can be modeled using the value of C_p in the glassy state and C_p in the rubbery state, as shown in Eq. (14).

$$C_{p-glass\ transition}(T) = C_{p-glass}(T) \cdot (1 - \mu(T)) + \mu(T) \cdot C_{p-rubbery}(T) \tag{14}$$

$\mu(T)$ is expressed by Eq. (15).

$$\mu(T) = \frac{1}{2}\left(1 + tanh\left(\frac{d}{c} \cdot T + e\right)\right) \tag{15}$$

where d is the temperature range of the glass transition period. c and e can be determined through cure fittings of the experimental data.

4.2.2 Thermal conductivity coefficient

Like the specific heat capacity, thermal conductivity coefficients of composites along fiber direction can be calculated with the mixing law. Thermal conductivity coefficients of composites in the crosswise fiber direction can be calculated using a composite law with Springer-Tsai model, as shown in Eq. (16).

$$K_{composite} = K_{epoxy} \cdot \frac{(1 + v_f\%) \cdot K_{fiber} + (1 - v_f\%) \cdot K_{epoxy}}{(1 + v_f\%) \cdot K_{fiber} + (1 + v_f\%) \cdot K_{epoxy}} \tag{16}$$

where $v_f\%$ represents the volume fraction of fiber. Values of solidified composite parts can be verified by thermal conductivity meters, such as the TA instrument LaserComp Fox314.

Thermal conductivity of epoxy also depends on temperature and degree of curing, especially unfully cured epoxy. For example, Bailleul et al. [14] found the thermal conductivity of raw material increased proportionally over temperature, while fully cured resin varied small over temperature.

Thermal conductivity of unfully cured epoxy can be measured by TA instrument DXF 500/900 models. Like specific heat capacity testing profiles, raw epoxy should be tested at several temperatures below its reaction initiation temperature, and fully cured epoxy are tested at temperatures below and above the T_g.

For thermal conductivity tests on solid material, guarded hot plate method is usually applied. Heat is applied on the top or bottom surface of a sample, and temperatures on both of the top and bottom surfaces are measured to calculate the thermal conductivity coefficient by Eq. (17).

$$K = \frac{qL}{T_H - T_L} \tag{17}$$

where q is the quantity of heat passing through a unit area of the sample in a unit time, with the unit of [W/m^2], L is the distance between the top and bottom surfaces of the sample with the unit of [m], T_H represents the temperature of the warmer surface with the unit of [K], and T_L represents the temperature of the cooler surface with the same unit. This method guarantees an accuracy of 98%, but it limits to low conductivity material, and it requires long measurement time and large specimen size [15]. It's often used to verify simulation values of solid FRC.

For liquid resin, hot wire or laser flash methods are usually applied. In a hot wire test, a heated wire is inserted into the material. Heat that flows out and temperature change in the wire are recorded to calculate the thermal conductivity. It's fast and with an accuracy up to 99% [15]. Laser flash method shoots short pulse of heat to the top surface of the sample, and temperature change is observed by an infrared scanner. This method requires small samples. It's fast and accurate at high temperatures but the instrument is expensive [15].

4.2.3 Resin enthalpy

The element that complexed the heat transfer calculation of thermoset composite is the reaction heat of unfully cured resin. The overall reaction enthalpy can be determined through a temperature scanning DSC test by integrating the heat flux over temperature. Enthalpy of epoxies that create different bonds during curing can vary dramatically. For example, a pre-catalyzed nadic methyl anhydride yields an enthalpy around 270 J/g in the reaction with a cycloaliphatic epoxy resin, while a formulated amine hardener yields a enthalpy around 410 J/g when reacted with a multifunctional epoxy resin.

Instantaneous reaction heat of an epoxy depends on degree of cure, curing rate, temperature, and fiber volume fraction. It can be represented by Eq. (18).

$$\dot{Q}_{reaction} = \frac{da}{dt}\left(1 - v_f\%\right) \cdot \rho_{resin} \cdot H_{enthalpy} \tag{18}$$

where a represents the degree of cure, $v_f\%$ is the fiber volume fraction, $H_{enthalpy}$ is the enthalpy of the epoxy, and ρ_{resin} is the density of the epoxy.

Like the density of water changes as it transfers to icy state, it's easy to imagine that the density of epoxy changes during its curing process, transferring from liquid to gelation and from glassy state to rubbery state. However, in the modeling of epoxy FRC, a constant epoxy density is usually applied, because (1) the existing of reinforcements lessened the flexibility of the volume change of epoxy and (2) the density change of epoxy is small. For example, the density change of MY750 epoxy is around 1% from unreacted state to fully cured state [16]. A density of 1.2 g/cm^3 is usually applied for epoxy. The density of composites can be calculated through the mix law.

With above illustrated parameters measured, the overall heat transfer governing equation can be used to model temperature histories of TSC at every locations of interest, as shown in Eq. (19). Thus, heat overshoot temperatures and locations can be predicted with the established model.

$$\begin{aligned}\frac{\partial}{\partial t}\left(\rho_{composite} C_{p-composite} T\right) &= \frac{\partial}{\partial x}\left(K_{xx-composite}\frac{\partial T}{\partial x}\right) + \frac{\partial}{\partial y}\left(K_{yy-composite}\frac{\partial T}{\partial y}\right) \\ &\quad + \frac{\partial}{\partial z}\left(K_{zz-composite}\frac{\partial T}{\partial z}\right) + \dot{Q}_{reaction}\end{aligned} \tag{19}$$

4.3 Residual stress characterization

Residual stress is introduced in epoxy TSC due to the curing shrinkage of epoxy and the mismatch between thermal expansion coefficient of the resin and reinforcements. Residual stresses, especially of TSC, are unevenly distributed in the composites. This is caused by the mismatch between composite layers and temperature/curing gradients.

Curing shrinkage of composites can be calculated by Eq. (20) [11], where H is a hindrance factor and can be found experimentally.

$$CS_{composite} = CS_{resin} - v_f\% \cdot H \tag{20}$$

Curing shrinkage of unreinforced epoxy can be tested by rheology method [17, 18] or gravimetric method [16, 18]. Rheometers are usually used in the rheology method. Two heated parallel plates are used to measure the linear shrinkage of the resin, which can be transferred into volume shrinkage, with assumptions that resin is incompressible and zero in-plane strains. Resin shrinkage before gelation cannot be measured with this method. The gravimetric method is based on the Archimedean principle, with which, density of an object can be measured by immersed in a fluid. This method involves immersing a silicon rubber bag with the testing resin in a known fluid and measuring the density of the resin while curing it. This method is able to obtain the curing shrinkage information both before and after resin gelation. A linear relationship was found between the curing shrinkage and degree of cure by Shah and Schubel [17] using the rheology method and Khoun and Hubert [18] found a bilinear relation in their study using the gravimetric method, where the rate changed at gelation.

The CTE of composite can be obtained through micromechanics model, shown in Eq. (21) for the fiber direction and Eq. (22) for transverse and thickness directions.

$$a_1 = \frac{a_{1f}E_{1f}v_f\% + a_{1r}E_{1r}(1 - v_f\%)}{E_{1f}v_f\% + E_{1r}(1 - v_f\%)} \tag{21}$$

$$a_2 = a_3 = (a_{2f} + v_{12f}a_{1f})v_f\% + (a_{2r} + v_{12r}a_{1r})(1 - v_f\%) - [v_{12f}v_{f\%} + v_{12r}(1 - v_f\%)]a_1 \tag{22}$$

where the subscript f represents reinforcements, while r represents resin. The subscript 1 represents the fiber direction (longitudinal) and 2 represents the transverse direction. E is tensile modulus and v_{12} is Poisson's ratio.

Young's modulus and Poisson's ratio can be measured using stain gage and Instron machine at various temperatures [9]. White and Hahn [19] found that both have a linear relationship to the degree of cure at a constant temperature. In their study, it showed that the CTE remained relatively constant in the transverse direction and its change in the longitudinal direction was negligible compared to the effect of curing shrinkage. Thus, in many studies, constant CTEs are applied for the resin and reinforcements [20].

5. Recommendations

In TSC manufacturing industry, micro-crack issue is the first and foremost important. Micro-cracks dramatically decrease the uniformity of mechanical properties of the TSC parts, especially shear strengths along composite plies. One way to predict micro-cracks is to analyze the TSC manufacturing process in two scenarios: (1) treating it as a flow issue as the boundaries gel but the core flows and (2) treating it as a stress issue when no resin flows in the TSC part. For example, a 6-inch-thick 12 inch by 12 inch epoxy TSC part is manufactured by Vacuum Resin Infusion Process. The epoxy gels 1 hour after mixing resin and hardener at room temperature. It takes 30 min to fully fill the entire part. After infusion, the part is placed in a convection oven for cure with a pre-subscribed temperature profile.

Due to the conventional heat inside the oven, resin at the outer surfaces of the part may solidify before the core curing to solid. In this scenario, the volume of the part is fixed by the solidified boundaries. Curing shrinkage of the core is the major cause of internal stress. The simulation can be simplified by separating the part into boundary (parts with gelled resin) and core (parts with ungelled resin). Gelation can be represented by degree of cure, which can be either simulated or tested by DEA. The shrinkage volume of the core after gelation can be simulated and compared to the current volume of the core. The total resin weight using to simulate the shrinkage volume of the core after gelation should be the total volume of resin infused into the part minus resin in the boundaries. A smaller shrinkage volume tends to result in micro-cracks at locations between layers in the core. This scenario is usually improved through the following methods:

- Infusing excess resin in the part through proprietary methods.

- Changing the solidification pattern of the part, e.g., B-stage the part before heating it up, or cool top surface, and/or heat bottom surface. For filament wound TSC billets, heating from inside out can largely reduce the chance of micro-cracks.

In the second scenario, after the part has no resin flowing, stresses develop between resin and reinforcements as well as among composite plies, due to degree of cure gradients. In the main author's earlier research [6], there was a 25% difference on degree of cure between the bottom and middle of the part. Internal stress in TSC parts at critical temperatures and degree of cures can be modeled and compared to the instantaneous strength of the composite. Larger stress predicts a high chance of micro-cracks. This scenario can be improved by adjusting curing profiles to avoid severe internal stress being built. B-stage and slower heating/cooling rate is usually helpful.

Acknowledgements

The support and motivation from Yanan and Liguo's family, their mom/mother-in-law, Shengmin Zhang, and their children, Leo Li and Landon Li, are greatly appreciated while Yanan and Liguo are writing this book chapter.

Author details

Yanan Hou[1]*, Liguo Li[1] and Joseph H. Koo[2]

1 The WellBoss Company LLC, Houston, TX, USA

2 Department of Mechanical Engineering, The University of Texas at Austin, Austin, TX, USA

*Address all correspondence to: yanan.hou@thewellboss.com

References

[1] Brosius D. An ode to the A380. Composites World. 2019;**5**:6

[2] Virginia Tech. Chemistry Researchers Use Block Copolymers to Create First Carbon Fibers with Uniform Porous Structure [Internet]. Available from: https://vtnews.vt.edu/articles/2019/02/mii-using-block-copolymers-to-make-porous-carbon-fibers.html#:~:targetText=Using%20block%20copolymers%20to%20create%20porous%20carbon%20fibers,-Liu%20used%20a&targetText=PAN%20is%20well%2Dknown%20in,removed%20to%20create%20the%20pores. [Accessed: December 11, 2019]

[3] CompositesWorld. GWEC Reports 51.3 GW of New Wind Capacity in 2018 [Internet]. Available from: https://www.compositesworld.com/news/gwec-reports-513-gw-of-new-wind-capacity-in-2018. [Accessed: December 11, 2019]

[4] Mason K. Spar forming simplified. Composites World. 2019;**5**:30-35

[5] Spencer JB. Cure Shrinkage in Casting Resins. United States: N.P.; 2015. DOI: 10.2172/1170250

[6] Hou Y. Vacuum-assisted resin infusion molding (VARIM) processing for thick-section composite laminates in wind turbine blades. In: Proceedings of the American Society for Composites 27th Technical Conference; 1–3 October 2012; Arlington. ASC Proceedings. 2012. pp. 334-351

[7] Li L. Interlaminar fatigue of thick-section glass fabric/vinyl ester composite [thesis]. Houston: University of Houston; 2013

[8] Makeev A, Nikishkov Y, Carpentier P, Lee E, Noel J. Manufacturing issues and measurement techniques for assessment of the effects on structural performance of composite parts. In: Proceedings of the American Helicopter Society 66th Annual Forum; 11–13 May 2010; Phoenix. 2010

[9] Griffin J. How to Dispose of 2-Part Epoxy Solutions [Internet]. 2013. Available from: https://www.lion.com/lion-news/april-2013/how-to-dispose-of-2-part-epoxy-solutions. [Accessed: October 16, 2019]

[10] Bailleul JL. Optimisation du cycle du caisson de pièces épaisses en matériau composite, Application a un préimprégné résine époxide/fibres de verre [thesis]. Nantes: Université de Nantes; 1997

[11] Nawab Y, Tardif X, Boyard N, Sobotka V, Casari P, Jacquemin F. Determination and modelling of the cure shrinkage of epoxy vinylester resin and associated composites by considering thermal gradients. Composites Science and Technology. 2012;**73**:81-87. DOI: 10.1016/j.compscitech.2012.09.018

[12] Hou Y, Wang SS. Analysis and Correlation of Curing Kinetics in Glass/Polyester Composite Processing by DSC and DEA Methods [Technical Report]. Houston: National Wind Energy Center; 2012

[13] Pascault JP, Sautereau H, Verdu J, Williams RJJ. Thermosetting Polymers. 1st ed. Boca Raton: CRC Press; 2002. 496 p. DOI: 10.1201/9780203908402

[14] Bailleul JL, Delaunay D, Jarny Y. Determination of temperature variable properties of composite materials: Methodology and experimental results. Journal of Reinforced Plastics and Composites. 1996;**15**:479-496. DOI: 10.1177/073168449601500503

[15] Buck W, Rudtsch S. Thermal properties. In: Czichos H, Saito T,

Smith L, editors. Springer Handbook of Materials Measurement Methods. Heidelberg: Springer Berlin Heidelberg; 2006. pp. 399-429. DOI: 10.1007/978-3-540-30300-8_8

[16] Li C, Potter K, Wisnom MR, Stringer G. In-situ measurement of chemical shrinkage of MY750 epoxy resin by a novel gravimetric method. Composites Science and Technology. 2004;**64**:55-64. DOI: 10.1016/S0266-3538(03)00199-4

[17] Shah DU, Schubel P. Evaluation of cure shrinkage measurement techniques for thermosetting resins. Polymer Testing. 2010;**29**:629-639. DOI: 10.1016/j.polymertesting.2010.05.001

[18] Khoun L, Hubert P. Cure shrinkage characterization of an epoxy resin system by two in situ measurement methods. Polymer Composites. 2010;**31**: 1603-1610. DOI: 10.1002/pc.20949

[19] White SR, Hahn HT. Process modeling of composite materials: Residual stress development during cure. Part II. Experimental validation. Journal of Composite Materials. 1992;**26**: 2423-2453. DOI: 10.1177/002199839202601605

[20] Bogetti TA, Gillespie JW Jr. Process-induced stress and deformation in thick-section thermoset composite laminates. Journal of Composite Materials. 1992;**26**: 626-660. DOI: 10.1177/002199839202600502

Chapter 7

Study of Composite Structures Based on a Porous Silicon Matrix and Nanoparticles Ag/Zno Used as Non-Invasive Highly Sensitive Biosensor Devices

Veniamin Koshevoi, Anton Belorus, Ilya Pleshanov, Anton Timchenko, Roman Denisenko, Daniyar Sherimov and Ekaterina Vodkailo

Abstract

In this work composite structures based on a porous silicon were obtained and studied. Porous matrices were formed by electrochemical etching in aqueous solutions of hydrofluoric acid. Based on the obtained substrates, por-silicon (Si)/ silver (Ag) and por-Si/zinc oxide (ZnO) composite structures were formed. These composites were functionalized by various methods (electro (E)-, thermo (T)-, electrothermal exposure) as a result of which the structures were modified. When studying the samples by scanning electron microscopy (SEM), it was concluded that silver nanoparticles actively diffused into the pores under these technological modes of functionalization. The por-Si/Ag and por-Si/ZnO composite structures were also studied using the following methods: infrared (IR) spectroscopy and Raman ultrasoft X-ray emission spectroscopy. Also, the photoluminescent characteristics of the samples were studied. Based on the obtained results, it was concluded that functionalization methods actively change the phase composition of structures and the optical properties of composites.

Keywords: porous silicon, composite structures, functionalization, nanoparticles, diffusion, phase composition

1. Introduction

One of the promising areas of science is biomedicine, which currently needs new materials and structures with certain requirements: biocompatibility, biodegradability, low toxicity, and the ability to use various methods of functionalization (post-processing) for a wide range of biologically active materials. One of the approaches to the formation of new structures is the creation of porous biocompatible matrices with deposited active particles. Due to their multifunctionality, these multilevel composite structures can be actively used as biosensors of a wide range of analytes, devices for targeted delivery of drugs, etc. [1–6].

The formation of various types of biosensors depends on the choice of material, the morphology of the sensitive layer, as well as its structural and physic-chemical properties. Various materials, such as Zn, Au, Ag, Pd, Pt, etc., as well as their modifications and composites can be used as an active layer. As an example, we can specify sensors based on various types of nanoparticles and composite structures: biosensors with zinc oxide (ZnO) nanoparticles, gold nanoparticles (AuNPs), Ag/TiO_2 nanoparticles, carbon nanotube-based biosensors, graphene quantum dot (GQD) biosensors, sensor based on the Pd/WO_3-ZnO composite porous thin films, etc. [6–17]. **Table 1** shows the sensitivity data of various composite sensor structures during glucose detection.

As already mentioned, changing the morphology and structure of the active layer also changes its sensitivity. **Table 2** shows an example of a change in sensitivity (glucose) with a change in the morphology of ZnO-based nanostructures [18].

As for the sensitive layer, the choice of the biosensor substrate material is an important factor. The choice of material as a porous silicon for the matrix is due to a wide range of key qualities required by structures for an effective use in biomedicine. These are the simplicity for functionalization, the ability to control the morphology and surface composition of porous matrices, the low cost of production, functional "flexibility," and good compatibility with the current industrial technologies. As a result, the porous silicon matrix can serve as the basis for composite structures. High values of surface area and specific area allow functionalization using a variety of biocompatible materials (silver, gold, magnetic metals, iron groups, tin, indium, zinc and their oxides).

The development of highly sensitive blood glucose meters is an extremely important task. This is especially important in the direction of personalized medical devices for people with diabetes. A promising task is the creation of noninvasive highly sensitive sensors with a low detection limit (of the order of 0.1 μM or less) [19–21]. The lower the detection limit parameter, the earlier a change in glucose level can be detected. These detection limit values can be achieved by using porous semiconductors as a substrate. It should be noticed that the morphological features of porous structures make it difficult to create a high-quality electrode base. To solve this problem, new methods have been developed for the functionalization of porous matrices with the aim of active deposition of nanoparticles into pores. The developed methods are non-destructive in comparison with analogues. The creation of a high-quality electrode base, with a large specific area of the sensitive layer, a high sensitivity parameter, a small detection limit, and small device sizes, is key to creating new personalized glucometers for recording glucose changes in direct time.

Thus, this work, the purpose of which is to study the physicochemical, electrical, and optical properties, as well as the morphological features of porous

Sensor type	Sensitivity
Glucose sensor based on Pd/WO_3-ZnO composite porous thin films	11.4 $\mu A\ \mu M^{-1}\ cm^{-2}$
Glucose biosensors with AuNPs (glassy carbon electrode (GCE) + AuNPs)	3.1 $\mu A\ mM^{-1}\ cm^{-2}$
Glucose biosensors based on GQDs	0.085 $\mu A\ \mu M^{-1}\ cm^{-2}$
Glucose biosensors based on Ag/TiO2 composite	39 $\mu A\ mM^{-1}\ cm^{-2}$
Glucose biosensors based on carbon nanotube	1433 $\mu A\ mM^{-1}\cdot cm^{-2}$
Glucose biosensors based on ZnO nanoparticles	15 $mA\ mM^{-1}\cdot cm^{-2}$

Table 1.
Sensitivity data for various types of biosensors during glucose recording [6–17].

Type of structures	Sensitivity
ZnO nanowire	26.3 mA mM^{-1} cm^{-2}
ZnO nanocomb	15.33 μA mM^{-1} cm^{-2}
ZnO:Co nanocluster	13.3 μA mM^{-1} cm^{-2}
Pyramid-shaped porous ZnO	237.8 μA mM^{-1} cm^{2}

Table 2.
Sensitivity of ZnO nanostructures with different morphology [18].

silicon matrices with varying technological parameters for the preparation of substrates and methods for their functionalization, is relevant and is of scientific and practical interest.

2. The formation of porous silicon matrices by electrochemical etching under various conditions

2.1 Method for the formation of porous silicon matrices

For the formation of a porous matrix, the method of electrochemical etching in aqueous solutions of hydrofluoric acid was used. A detailed installation diagram is presented in **Figure 1**. The structure of the stand can be divided into separate components: a glassy carbon crucible which contains electrolyte (1), a silicon wafer (2), and devices used to move the crucible and the sample in different planes (3)–(5).

During the electrochemical process, silicon atoms actively react with the molecules (ions) of the electrolyte when they come into solution. This process was formulated by Lehman and Joselle, describing the migration of holes at the surface of a sample when a current is applied. Therefore, it is worth considering that such etching modes as current density and anodization time directly affect the structure of the porous layer. The composition of the electrolyte also affects the morphology of the matrix. In this work an aqueous solution of hydrofluoric acid with the

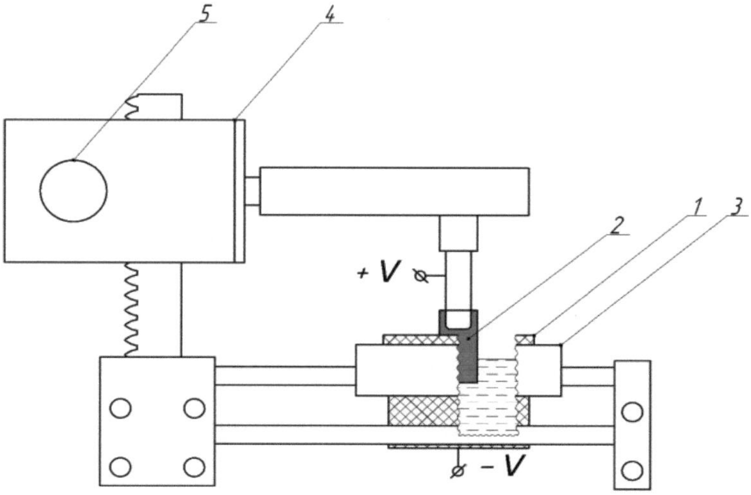

Figure 1.
Installation for electrochemical etching: (1) crucible; (2) sample; (3) platform; (4) carriage; and (5) height adjustment knob.

addition of isopropyl alcohol (C_3H_7OH) was used. This solution provides the necessary wettability of the silicon wafer surface by electrolyte [22–25].

As a result of electrochemical etching, observing the galvanostatic mode (current density j = 30 mA/cm^2: anodizing time t = 10 min), porous substrates were formed. New structures were subsequently used as the basis for the formation of por-Si/Ag and por-Si/ZnO composite structures.

2.2 Functionalization methods

To create functionalized composite structures, Ag nanoparticles were deposited on the formed porous substrates. Also, under an electrothermal influence, it was possible to ensure the diffusion of nanocrystals into the matrix. These functionalization methods allow the formation of composites with new structural and physicochemical properties.

2.2.1 Synthesis of silver nanoparticles using silver nitrate ($AgNO_3$)

In this work, silver ink using the recovery method with sodium citrate and with the addition of a strong reducing agent, which allows one to achieve the necessary concentration of supersaturation in the early stages of synthesis, was obtained [26]. Ascorbic or tannic acid can act as a strong reducing agent.

The solution is formed by heating for 30 min a mixture based on silver nitrate, sodium citrate and sodium chloride, and a solution of ascorbic acid heated to 95°C is added. Then, unreacted components are removed from the solution by centrifugation. Subsequently, in several stages, silver layers were applied followed by annealing at a temperature of 150°C for 30 min.

To determine the sizes of the synthesized silver nanoparticles, the scanning electron microscopy (SEM) method was used (**Figure 2**).

Based on the obtained images, it was concluded that the diameter of silver nanocrystals is about 20 nm.

2.2.2 Functionalization method based on the electrochemical deposition of zinc oxide nanoparticles

In this work, the functionalization of zinc oxide nanoparticles with porous silicon samples was carried out using the electrochemical deposition of metals from an electrolyte solution [27]. For this methodology an installation was developed as shown in **Figure 3**.

The electrodeposition was carried out as follows: in a crucible (6) with electrolyte (5), which was previously heated to a temperature of 65°C, a previously prepared sample of a porous silicon (4) was immersed. An electric current in the range from 0.5 to 1 A was passed through anode (6) and cathode (3), which provided the beginning of the electrodeposition process. Motor (2) provided a continuous mixing of the electrolyte solution during electrodeposition at a speed of 5000 r/s for 1 h. Then, the samples of porous silicon with a deposited layer of zinc were annealed in a muffle furnace at a temperature of 200°C for 20 min.

2.2.3 Functionalization method based on electrothermal diffusion processes

To ensure a high-quality introduction of silver nanoparticles into porous matrices, the method of electrothermal diffusion was used. Installation for the implementation of this process is presented in **Figure 4**.

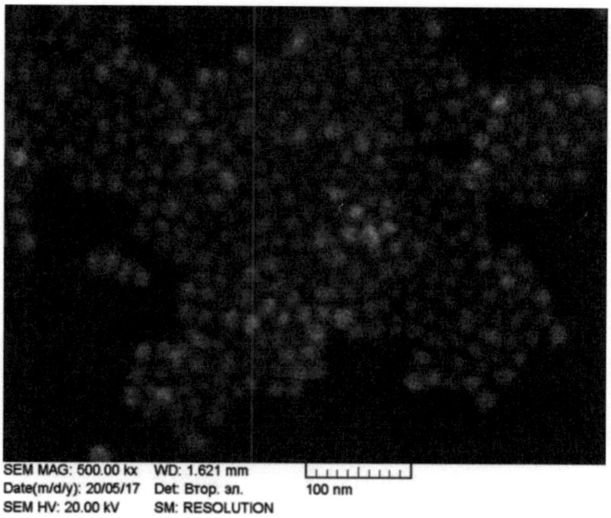

Figure 2.
Image of synthesized silver nanoparticles.

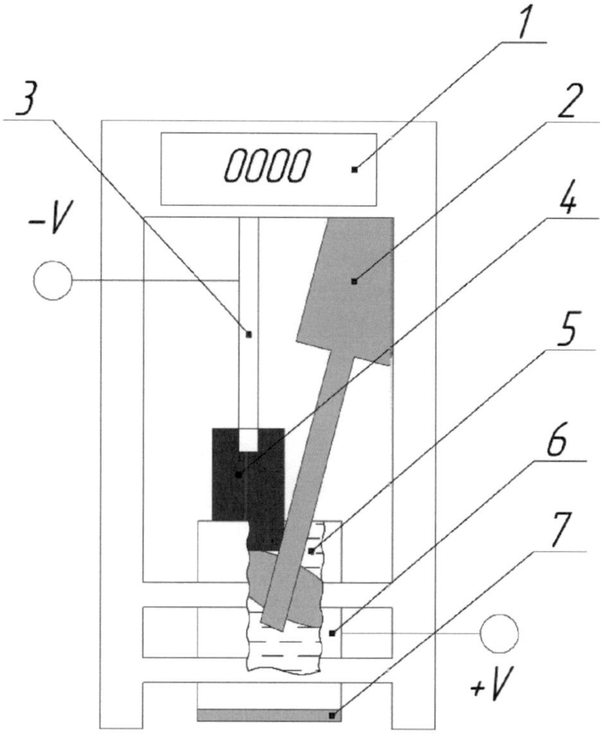

Figure 3.
Installation for electrodeposition of metals: (1) indicator of motor rotation speed; (2) motor; (3) cathode; (4) sample (to which electrodeposition is performed); (5) electrolyte; (6) graphite crucible (anode); and (7) heating element (Peltier element).

This installation includes a muffle furnace (1) (capable of maintaining a temperature of up to 900°C) with a fixed sample on the worktable. The clamping mechanism (acting as the anode) fixes the sample and implements the current to the surface layer.

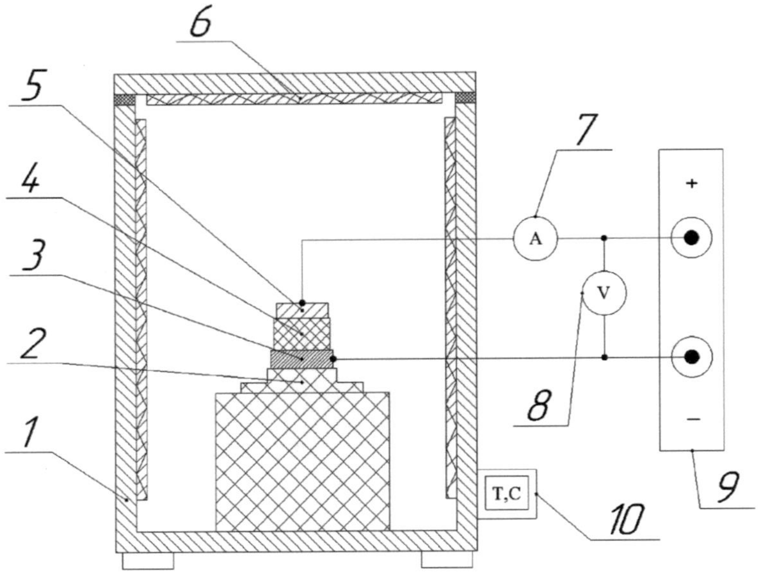

Figure 4.
Installation for functionalization by electro (E)-diffusion method: (1) muffle furnace; (2) desktop; (3) a metal film (cathode); (4) porous semiconductor; (5) metal layer (anode); (6) heaters; (7) milliammeter; (8) voltmeter; (9) power supply; and (10) thermometer.

Samples, on which silver layers were deposited by the method of colloidal quantum dots, were fixed on the worktable. Then, the table was placed in a muffle furnace and preheated to 200°C for 15 min. This process is necessary to avoid thermal shock and prevent the destruction of the material. To activate electro-diffusion, a voltage of 300 V was applied to the samples. This electrothermal process was observed for 25 min. Then, the voltage supply stopped and the samples are cooled for 30 min inside the furnace. Using this method a series of samples in which silver nanoparticles diffused deep into the porous matrix were obtained. The thermal deposition of metals into the porous layer took place at the same parameters as the deposition by electrothermal diffusion, but without applying an electric potential to the sample [28–30].

3. Assessment of the morphology of nanostructures using SEM methods

In this work, the samples formed during the experiments were studied by scanning electron microscopy (TESCAN MIRA3 electron microscope). This technique allowed us to determine the morphological features of porous matrices before and after functionalization processes. The measurement results are presented in **Figures 5–9**.

Based on the obtained images (**Figure 5**), it was concluded that the size of pores for por-Si matrices is about 50 nm and the thickness of the porous layer is 20 μm. **Figure 6** shows the images for porous silicon matrices after the functionalization process (deposition of silver nanoparticles by the method of colloidal quantum dots).

The layered deposition of silver nanoparticles allows to achieve good uniformity on the surface of the substrate. Also, according to the data obtained (**Figure 6**), it was found that silver is predominantly localized on the porous matrix and only a small fraction goes deep into the wide pores. **Figure 7** shows SEM images of samples coated with a layer of silver after the electro-diffusion process.

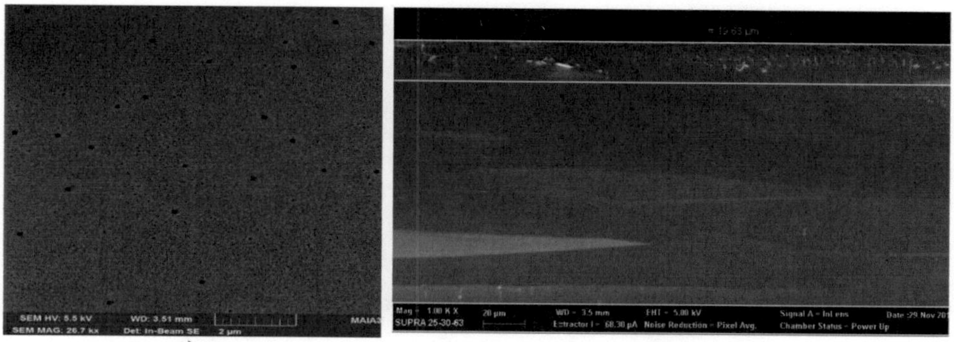

a) b)

Figure 5.
Scanning electron microscopy data of a porous silicon matrix: (a) surface and (b) cross section.

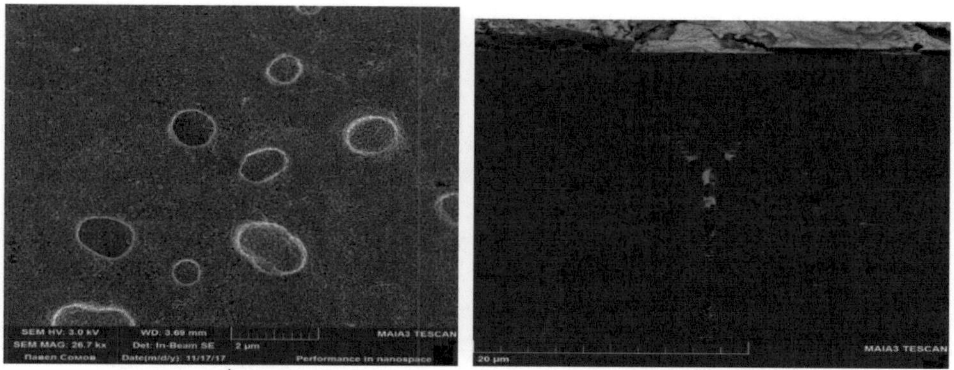

a) b)

Figure 6.
Scanning electron microscopy data of por-Si/Ag structures: (a) surface and (b) cross section.

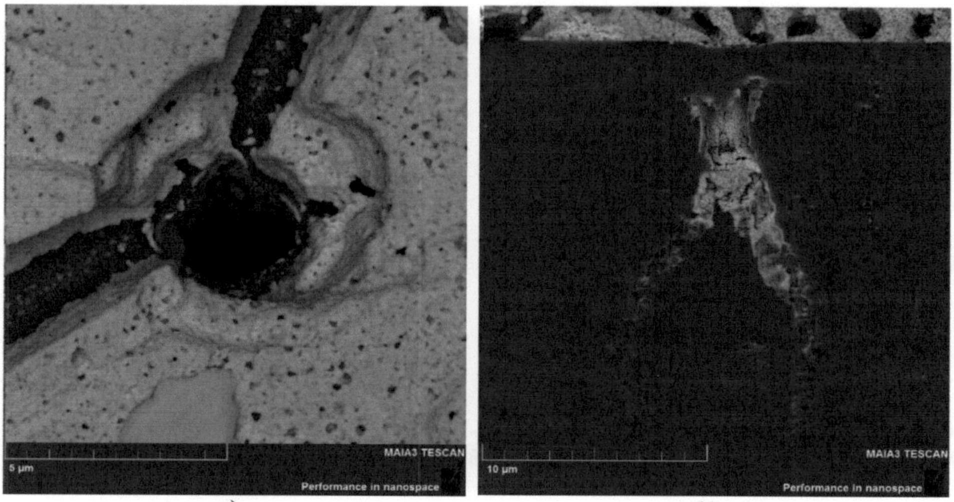

a) b)

Figure 7.
SEM data of por-Si/Ag structures after the process of electro-diffusion: (a) surface and (b) cross section.

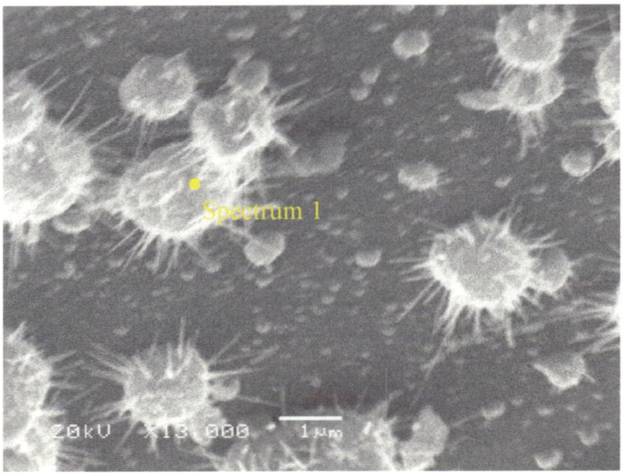

Figure 8.
SEM data of the por-SI/ZnO structure.

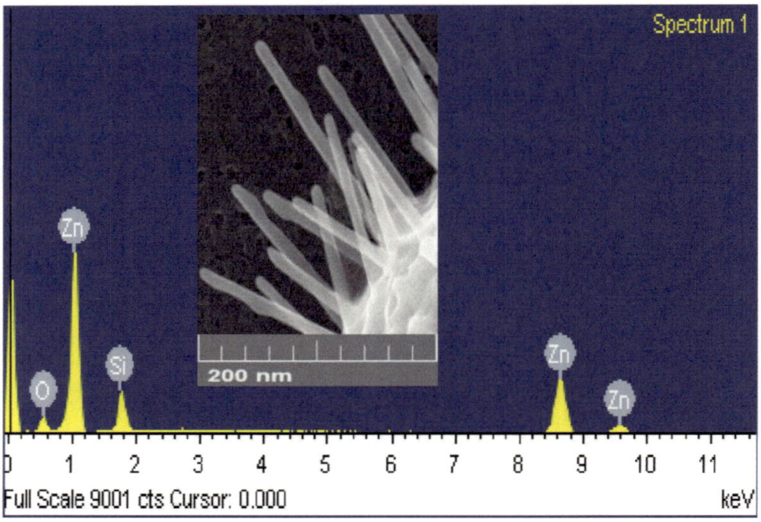

Figure 9.
EDX analysis of por-Si/ZnO structure.

It was found (**Figure 7**) that after the functionalization process based on electrodiffusion, silver actively diffuses into the pores. The combined effect of the electric and thermal conditions ensures the activation of this process. Unlike conventional layering, silver nanoparticles occupy almost the entire pore depth.

3.1 Study of composite structures por-Si/ZnO by SEM method

To study the por-Si/ZnO composite structures, scanning electron microscopy (TESCAN MIRA3 microscope) was used. Also energy-dispersive X-ray spectroscopy (EDX) to analyze the composition was used. The data are presented in **Figures 8** and **9**.

In **Table 3** the EDX analysis data described in **Figure 9** is presented.

Thus, it was found that zinc particles predominate in the studied structures. The mass fraction of Zn is 77.83% of the total mass of por-Si/ZnO. It can be argued that

Element	Mass, %	Atom content, %
O K	10.51	29.03
Si K	11.66	18.35
Zn K	77.83	52.62
Total	100	100

Table 3.
EDX data analysis.

during the electrochemical deposition of zinc oxide nanoparticles, the active passivation of the porous surface occurs, followed by the oxidation of the nanoparticles. Also based on SEM data, it was concluded that the largest ZnO nanoparticle size reaches about 2 μm. The size of the rods in this case varies in the range from 200 nm to 1 μm. These particles are formed on the surface of the porous matrix, fixing on its surface.

4. The study of composite structures by IR and Raman spectroscopy

4.1 The study of composite structures by IR spectroscopy

The IR transmission spectra of the samples were obtained on a Vertex 70 FTIR spectrometer (Bruker) using an attachment for spectroscopy of impaired total internal reflection (TIR) [31]. The depth of study of porous silicon samples by wavelengths up to 2000 cm^{-1} does not exceed 1.5 μm by this technique, and in the range of 2000–4000 cm^{-1}, it does not exceed 10 μm. Thus, based on our calculations and the data on the thickness of the porous layers obtained using SEM, by the ATR method, we obtain data mainly on the composition of the porous layer with an insignificant contribution from the substrate. IR spectra were obtained a month after the manufacture of the samples (**Figure 10**).

According to IR spectroscopy, after 30 days of exposure to the atmosphere, the spectrum of a "standard" por-Si sample shows bands characteristic of porous silicon, corresponding to Si-Si and Si-H bonds (616 and 634 cm^{-1}); band 750–1000 cm^{-1}, corresponding to different types of Si-H and O_xSi_y bonds; the oxide compound peak of Si-O-Si (1060–1250 cm^{-1}); and the band 2150–2250 cm^{-1}, corresponding to various types of vibration of bonds of adsorbed hydrogen and oxygen-containing groups [32]. In addition, the band at 480 cm^{-1} of the spectrum corresponds to deformation vibrations of Si-O-Si.

The spectra of porous silicon samples with deposited silver nanoparticles were uninformative, apparently, due to the high reflectivity of the formed silver film. No pronounced absorption bands were observed on the spectra of these samples by this technique.

The situation is different for por-Si samples with various methods of functionalization (electro-, thermo-, and electrothermodiffusion methods) of ZnO films. The deposition of zinc oxide on the surface leads to a change in the degree of oxidation of the surface, a decrease in the intensity of the band of the oxide composite peak of Si-O-Si stretching vibrations (1060–1250 cm^{-1}), and a characteristic of stoichiometric silicon dioxide (SiO_2) with an increase in the intensity of the absorption peak corresponding to Si-O deformation vibrations—Si. The band corresponding to Si-Si bonds also becomes less pronounced against the background of an increase

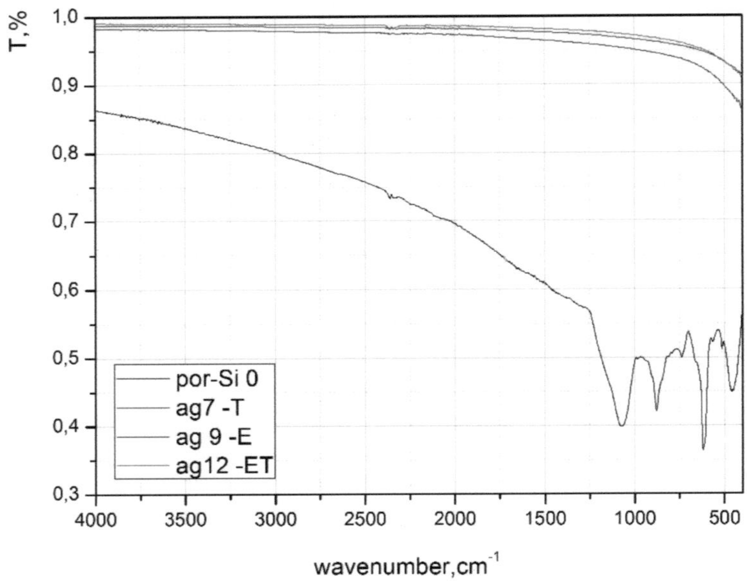

Figure 10.
IR transmission spectra of samples of porous silicon with deposited silver functionalized by electro-, thermo-, and electrothermodiffusion (E-T) methods.

in the intensity of the complex compound band of 750–1000 cm^{-1}, corresponding to various types of Si-H and O_xSi_y bonds. A change in the shape of the absorption band of the spectrum of 750–1000 cm^{-1} shows that the deposition of zinc oxide leads to a decrease in the fraction of Si-H_x bonds on the surface of the porous layer and a corresponding increase in oxygen-containing bonds of the O_xSiH_y type. Comparing the samples of porous silicon with ZnO functionalized by different methods, we can conclude that, with the general tendency toward a change in the oxidation state of porous silicon described above, the electrical effect of zinc oxide affects the por-Si surface to a lesser extent than thermal and electrothermal effects (**Figures 11** and **12**).

Comparing the samples of porous silicon with deposited ZnO functionalized by different methods, we can conclude that, with the general tendency toward a change in the oxidation state of porous silicon described above, the thermal deposition of zinc oxide affects the por-Si surface to a lesser extent compared to electrically and electrothermally modified.

4.2 The study of composite structures by Raman spectroscopy

Raman spectra (Raman light scattering) were recorded in the backscattering geometry at room temperature using a Raman Microscope RamMics M532 EnSpectr laboratory Raman spectrometer. As an excitation source, a 532 nm Nd:YAG laser was used. The Raman spectroscopy method was used as an addition to the IR spectroscopy method to determine the effect of metal film deposition on post-processing on the composition of the matrix of porous silicon. The estimated depth of analysis by this technique was about 1 μm.

The Raman spectrum of porous silicon has the same features as the spectrum of the crystalline silicon substrate on which it was grown. However, the spectrum of porous silicon broadened the main band corresponding to the TO-phonon line of silicon (520.7 cm^{-1}), as well as more pronounced features at 300 (LA phonon of silicon)

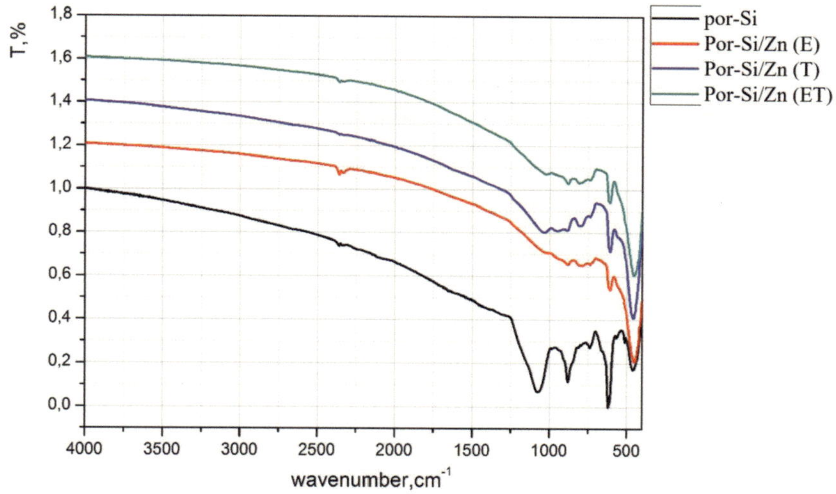

Figure 11.
IR transmission spectra of samples of porous silicon with zinc oxide layers functionalized by the methods of electro-, thermo-, and electrothermodiffusion methods.

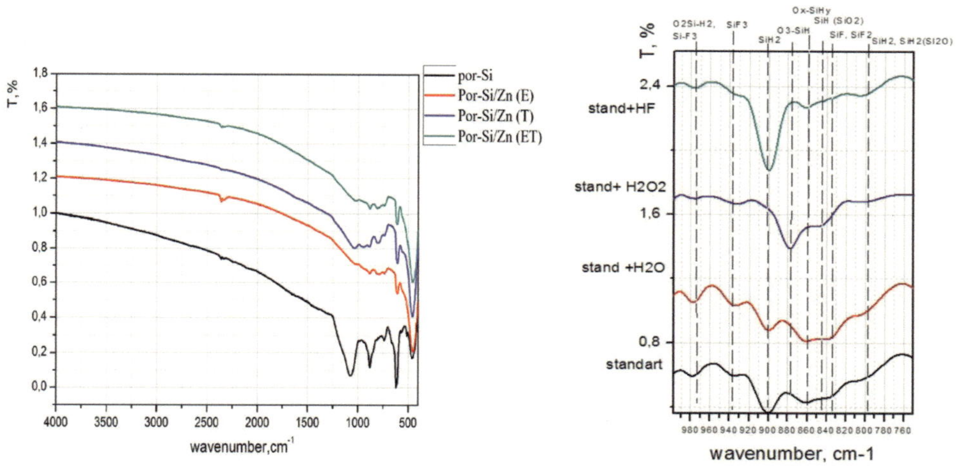

a) b)

Figure 12.
(a) IR transmission spectra of samples of porous silicon with deposited zinc oxide nanoparticles functionalized by electro-, thermo-, and electrothermodiffusion methods, the spectral region is 760–1500 cm^{-1}. (b) An example of deciphering the features of the IR spectrum of porous silicon samples in the region of 760–960 cm^{-1} according to [32].

and 950–1000 cm^{-1} (TO-phonon, second order). The broadening of the main band and the increase in the intensities of the minor bands of the Raman spectrum can be explained by the structural deformations of silicon during the formation of pores in it (**Figures 13** and **14**).

It was found that the deposition of silver and zinc oxide into porous silicon does not significantly affect the shape of the spectrum of porous silicon. In this case, however, a slight broadening of the main band of the TO-phonon of silicon occurs, which indicates a slight additional deformation of the porous matrix; moreover, a wide band of 1200–1600 cm^{-1} appears on the Raman spectra of the samples with deposited metals, which corresponds to carbon contamination of the surface by the deposition products (amorphous carbon).

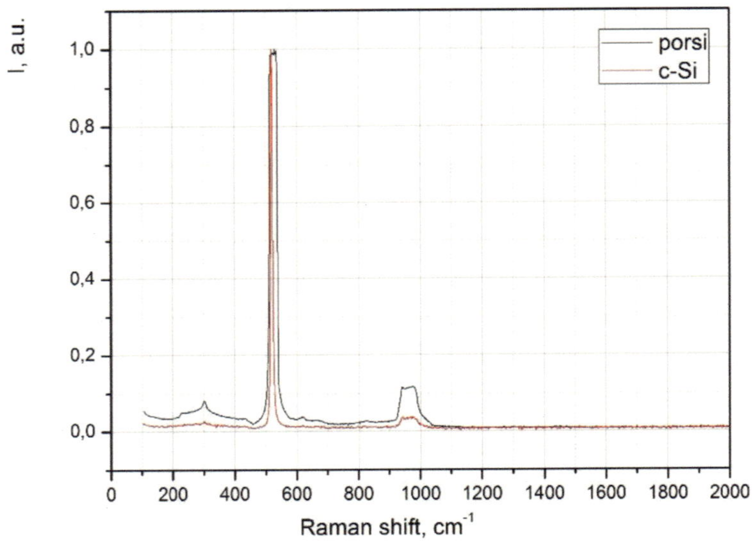

Figure 13.
Normalized Raman spectra of porous silicon samples and c-Si substrates.

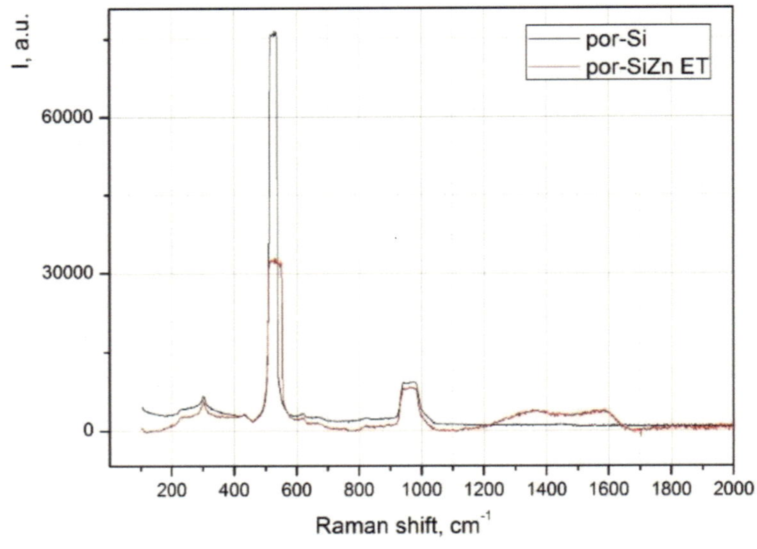

Figure 14.
Raman spectra of the samples por-Si and por-Si/Zn under an electrothermal diffusion effect.

5. The study of samples using an ultrasoft X-ray emission spectroscopy

X-ray emission USXES Si L2,3 spectra of the samples were obtained on a unique laboratory X-ray spectrometer-RSM-500 monochromator, in high vacuum (3×10^{-6} mmHg); the energy resolution of the obtained spectra was 0.3 eV. This technique is sensitive to the local partial density of states of atoms of a certain sort, due to which a qualitative assessment of the atomic structure of porous silicon and a semiquantitative assessment of the relative phase content in this material are possible [33]. In the case of porous silicon, USXES Si L 2,3 spectra reflect the distribution of Si 3S + Si 3d states with a predominance of Si 3 s. The depths of analysis by this method were ~30 and 60 nm at an accelerating voltage at the anode of the X-ray tube, on which the test sample was placed, 2 and 3 kV, respectively.

The obtained X-ray emission spectra were processed using a special computer program, which allows the phase composition of the obtained por-Si samples to be determined by summing the spectra of reference materials with the corresponding weight coefficients that can be part of the porous layer [33]. When simulating the Si L 2,3 spectra of por-Si samples, the following reference spectra were used: single-crystal silicon (c-Si); amorphous hydrogenated silicon (a-Si:H); low-coordinated (lc) silicon Si, which was observed in amorphous Si films [34] (coordination number ~2.5–3); disordered silicon after implantation with argon Si:Ar; silicon suboxide (SiOx), where x is 1.3; and SiO_2 [34, 35]. The modeling error was determined as the difference between the areas under the experimental and simulated Si L 2,3 spectra and did not exceed 10%. In addition the general form of the simulated and experimental spectra was taken into account [36]. The survey was carried out 1 month after receiving the samples (**Figure 15**).

Figures 16 and **17** show the USXES Si L2,3 spectra of samples of the initial porous silicon and porous silicon with deposited silver and zinc oxide nanoparticles, obtained for the analysis depths of 30 (a) and 60 (b) nm.

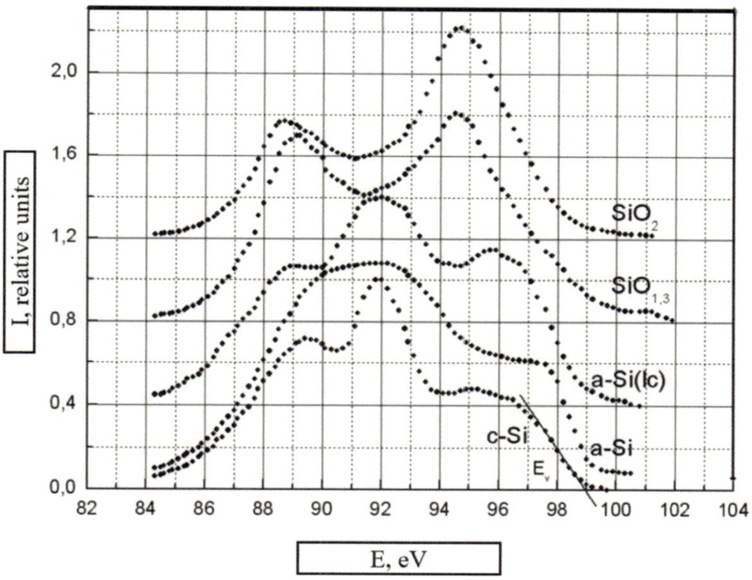

Figure 15.
USXES Si L2,3 spectra of reference phases.

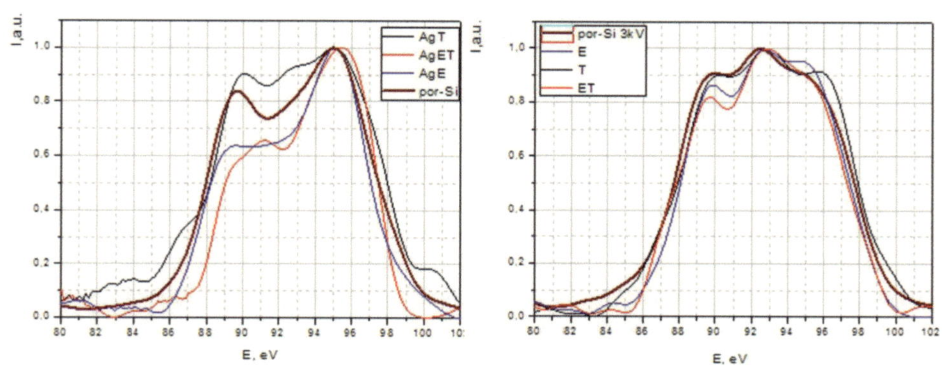

Figure 16.
USXES Si L2,3 porous silicon and por-Si with deposited silver nanoparticles.

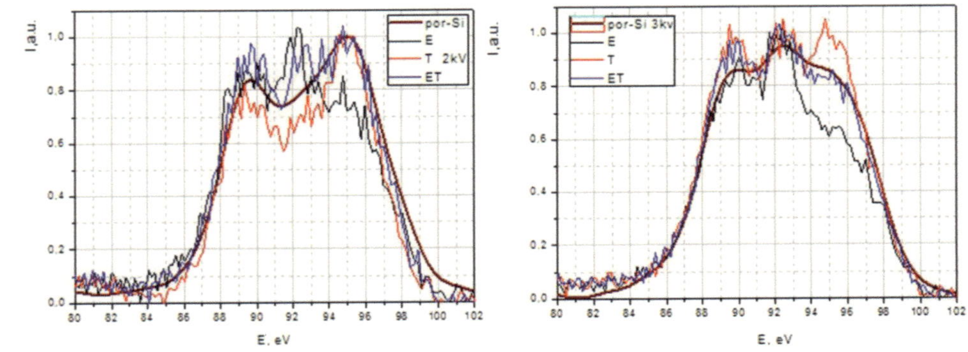

Figure 17.
USXES Si L2,3 porous silicon and por-Si with deposited ZnO nanoparticles.

At an analysis depth of 30 nm by the USXES method, the following data were obtained. The closer to the surface layer of the initial porous silicon is predominantly oxidized, the fraction of unoxidized silicon (c-Si, a-Si) on the surface is less than 50% (burst 10%), while por-Si/Ag substrate functionalized by the electrical and electrothermal methods leads to stronger oxidation matrices predominant in the surface composition of SiO_2 oxide. Thermal method of modification has a lesser effect on the surface composition of the porous silicon matrix used, while the fraction of unoxidized phases in the surface composition slightly increases. This can be attributed to the morphological features of the film grown on the surface by this technique.

With a greater depth of analysis by the USXES method, at 60 nm, the following data were obtained: the phase composition of more porous silicon is expectedly less oxidized compared to its surface; the contribution of crystalline and amorphous silicon phases prevails over the contribution from oxide phases. In this case, the por-Si/Ag samples functionalized by all methods have a significantly smaller effect on the composition of the por-Si matrix layer deeper; phase analysis shows approximately the same ratio of oxidized and non-oxidized phases as in the initial porous silicon with a slight increase in the oxidation state, i.e., defective SiOx is oxidized to SiO_2, while maintaining the proportion of SiOx + SiO_2 in the total phase composition.

The following data were obtained for composites with precipitated ZnO at an analysis depth of 30 nm using the USXES method. Recall that we used the same matrix of porous silicon in the same series of samples. As in the case of the deposition of silver nanoparticles, the functionalization of zinc oxide structures by the thermal method has the least effect on the surface composition of the porous silicon matrix used (toward a slightly higher matrix oxidation). This can be attributed to the morphological features of the film grown on the surface by this technique. The functionalization of ZnO particles by electro- and electrothermal diffusion methods leads to a stronger change in the composition of the matrix surface, namely, a significantly smaller contribution of silicon oxide phases, with the most pronounced changes observed in samples with ZnO electrochemical modification.

With a greater depth of analysis by the USXES method, at 60 nm, the following data were obtained: the most pronounced changes in the phase composition, as well as at a shorter depth of analysis, are observed in samples with por-Si/ZnO samples with electrical modification; the degree of oxidation of the porous layer in these samples is significantly lower than the initial porous silicon. In electrothermal and thermal functionalization, the most severe changes which are observed in por-Si/ZnO samples with electrical modification to a lesser extent affect the composition of the matrix, leading to a slight increase in the fraction of oxide phases.

6. The study of the photoluminescent characteristics of samples

The photoluminescent (PL) spectra of multilayer porous silicon were measured using an experimental setup based on an OceanOptics USB4000-VIS–NIR (350–1000 nm) optical fiber spectrometer. An HP Lightning light-emitting diode (LED) was used as the PL excitation source, the maximum radiation wavelength of which was 405 nm (~3.06 eV). A quartz-focusing lens and a band-pass filter were installed in front of the LED, cutting out the long-wavelength region of the LED's radiation. The PL radiation of the samples was introduced into the QP600–2-UV–VIS (OceanOptics) fiber through a collimator. Next, the radiation was directed to the input of the spectrometer. The measurements were carried out in the dark in the absence of scattered light sources a month after the preparation of the samples.

The photoluminescence spectrum of the initial porous silicon is a band in the region of 500–800 nm with a peak of ~650 nm and is a characteristic of por-Si samples obtained by this method. The deposition of silver nanoparticles using

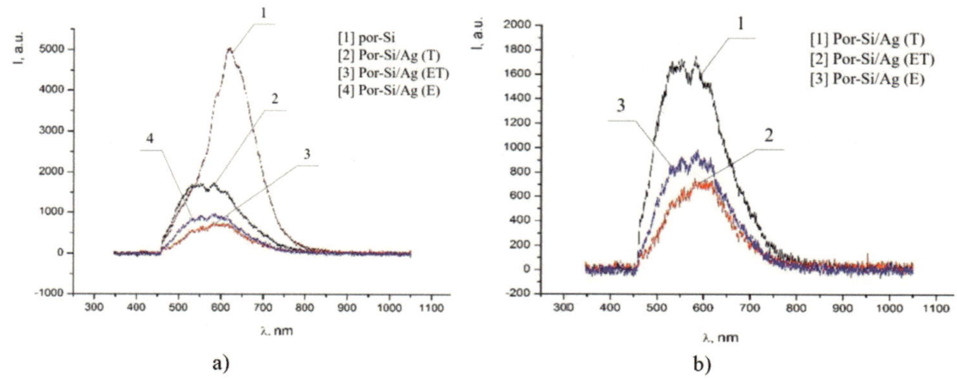

Figure 18.
PL spectra of samples of porous silicon and porous silicon with deposited silver nanoparticles with different types of functionalization; a) full spectra for all samples, b) local part of the spectrum: (por-Si) - porous silicon, (por-Si/Ag (T)) porous silicon with Ag nanoparticles functionalized by thermal exposure, (por-Si/Ag (E)) porous silicon with Ag nanoparticles functionalized by electrical exposure, (por-Si/Ag (ET)) porous silicon with Ag nanoparticles functionalized by electrothermal exposure.

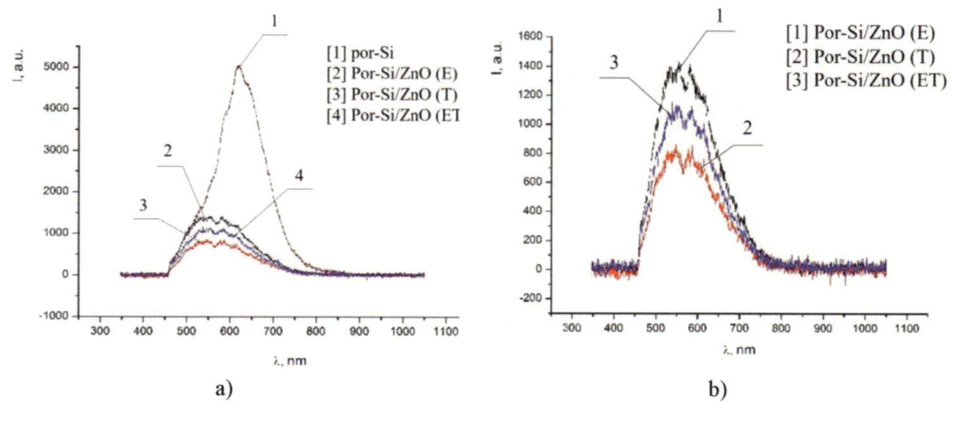

Figure 19.
PL spectra of samples of porous silicon and porous silicon with deposited ZnO nanoparticles with different types of functionalization; a) full spectra for all samples, b) local part of the spectrum: (por-Si) - porous silicon, (por-Si/ZnO (T)) porous silicon with ZnO nanoparticles functionalized by thermal exposure, (por-Si/ZnO (E)) porous silicon with ZnO nanoparticles functionalized by electrical exposure, (por-Si/ZnO (ET)) porous silicon with ZnO nanoparticles functionalized by electrothermal exposure.

the techniques used in this work leads to a noticeable shift of the PL peak toward shorter wavelengths with a slight decrease in the PL intensity. The peak of the PL of samples with por-Si/Ag samples modified by various techniques is in the region of 550–600 nm, with the highest PL intensity having the composite sample functionalized by thermal exposure (**Figures 18** and **19**).

The deposition of zinc oxide nanostructures by the methods used in this work, as in the case of deposition of silver nanoparticles, leads to a noticeable shift of the PL peak toward shorter wavelengths with a slight decrease in the PL intensity. The peak of the PL of por-Si/ZnO samples functionalized by various techniques is in the range of 550–600 nm, with the highest PL intensity having the composite sample modified by electrical exposure.

7. Conclusions

1. There were a series of porous silicon substrate by electrochemical etching obtained. To create functionalized composite structures, Ag and ZnO nanoparticles were deposited on the formed porous substrates. Silver ink was formed by reducing sodium citrate at high supersaturation concentrations. Zinc oxide nanoparticles were deposited on substrates by electrochemical deposition of metals from an electrolyte solution. To ensure the effective introduction of silver and zinc oxide nanoparticles into the porous matrix, the method of electrothermal diffusion was used.

2. There were obtained data by scanning electron microscopy for all series of samples. Based on the analyzed images, it was concluded that the porous silicon substrate pore size is about 50 nm and the thickness of the porous layer 20 μm. When silver nanoparticles are deposited on a porous matrix, localization is achieved mainly on the surface of the substrate. By the method of electrothermal diffusion of the nanoparticle, active diffusion of particles deep into the pores was achieved. Silver actively diffuses into the pores, occupying almost the entire pore depth. Thus, SEM made it possible to estimate the particle size of ZnO reaching up to 2 μm (rod lengths in the range from 200 nm to 1 μm). Particles of such sizes do not penetrate into the pores and remain fixed on the surface of the porous matrix.

3. There were obtained data on the surface composition of nanostructures by IR and Raman spectroscopies. The IR transmission spectra of the samples were detected on a Vertex 70 FTIR spectrometer (Bruker). An analysis of the spectra indicates that the exposure of the samples in the atmosphere activates the oxide compound peak Si-O-Si (1060–1250 cm^{-1}). Also, on the spectra there is a band at 480 cm^{-1} corresponding to deformation vibrations of Si-O-Si.

   Due to the high reflectivity of the silver layer, it was not possible to reveal pronounced absorption bands in the spectra. In this case, the obtained data for por-Si/ZnO composite structures led to the conclusion that the deposition of a zinc oxide layer on the substrate leads to a change in the oxidation state of the surface, which leads to a decrease in the intensity of the oxide composite peak band of Si-O-Si stretching vibrations (1060–1250 cm^{-1}). The spectra showed changes in the shape of the absorption band of the spectrum of 750–1000 cm^{-1}. This fact indicates a decrease in the fraction of Si-Hx bonds and a corresponding increase in oxygen-containing bonds of the OxSiHy type when silver nanoparticles are deposited on the substrate surface.

4. By the method of ultrasoft X-ray emission spectroscopy, data on the atomic and electronic structures and phase composition of the nanostructures were obtained. USXES X-ray emission spectra indicate that the porous matrix is primarily predominantly oxidized by a, the fraction of unoxidized silicon (c-Si, a-Si) on the surface is less than 50% (burst 10%), while the deposition of silver nanoparticles leads to stronger oxidation of the matrix, the prevalence of SiO_2 oxide in the surface composition. In this case, an increase in the fraction of unoxidized phases in the surface composition is observed.

   For composite structures based on por-Si/ZnO and por-Si/Ag, the USXES method was used to analyze the phase composition at various depths of 30 and 60 nm. For both types of composite matrices, temperature oxidation at depths of 30 nm and 60 nm results in the oxidation of the active surface, but the overall composition remains almost unchanged. It should be noted that during the functionalization of composite structures by the methods of electro- and electrothermodiffusion, a significant change in the composition of the surface layer occurs as a result of diffusion processes. A significant decrease in the oxide phases of silicon was also noted.

5. The photoluminescent characteristics of the samples were investigated using a USBO4000-VIS–NIR fiber-optic spectrometer (350–1000 nm) from Ocean-Optics (the maximum radiation wavelength of which was 405 nm (~3.06 eV)). When studying the spectra of substrates, characteristic spectra were obtained for this type of porous matrices (a band in the region of 500–800 nm with a peak of ~650 nm). The functionalization of the surface by silver nanoparticles leads to a noticeable shift of the PL peak toward shorter wavelengths with a slight decrease in the PL intensity. The electrical, thermal, and combined effects are determined on the PL spectra by a peak in the region of 550–600 nm, while the composite structure corresponding to the thermal effect has the highest PL intensity.

   The deposition of zinc oxide nanostructures by the methods used in this work, as in the case of deposition of silver nanoparticles, leads to a noticeable shift of the PL peak toward shorter wavelengths with a slight decrease in the PL intensity. The PL peak of samples with ZnO nanostructures deposited by various techniques is in the range of 550–600 nm. Unlike por-Si/structures, the highest PL intensity has a sample that was subjected to electrothermal interaction.

6. The main results of this work include the data obtained during the study, which allow to determine the patterns of influence of the presented functionalization modes on the characteristics of porous matrices. The studied structures can be used as sensory devices with an increased level of sensitivity and have a wide range of detectable analyses.

Author details

Veniamin Koshevoi[1]*, Anton Belorus[2], Ilya Pleshanov[3], Anton Timchenko[4], Roman Denisenko[3], Daniyar Sherimov[2] and Ekaterina Vodkailo[1]

1 St. Petersburg Mining University, Saint-Petersburg, Russia

2 St. Petersburg Electrotechnical University "LETI", Saint-Petersburg, Russia

3 St. Petersburg National Research University of Information Technologies, Saint-Petersburg, Russia

4 Immanuel Kant Baltic Federal University, Kaliningrad, Russia

*Address all correspondence to: venia.koshevoi.eltech@gmail.com

References

[1] Leigh C. Handbook of Porous Silicon. 2nd ed. Switzerland: Springer International Publishing; 2014. p. 1613. DOI: 10.1007/978-3-319-05744-6

[2] Yin J, Qi X, Yang L, Hao G, Li J, Zhong J. A hydrogen peroxide electrochemical sensor based on silver nanoparticles decorated silicon nanowire arrays. Electrochimica Acta. 2011;**56**:3884-3889. DOI: 10.1016/j.electacta.2011.02.033

[3] Martín-Palma RJ, McAtee PD, Ramadan R, Lakhtakia A. Hybrid nanostructured porous silicon-silver layers for wideband optical absorption. Scientific Reports. 2019;**9**(1):7291. DOI: 10.1038/s41598-019-43712-7

[4] Irina K, Mihaela M, Mihai D, Monica S, Teodora I, Adina B, et al. Silver/porous silicon (PS) nanocomposite layers for biomedical applications. In: Proceedings of International Semiconductor Conference; 27-29 September; Sinaia. Sinaia: IEEE; 2006. DOI: 10.1109/SMICND.2006.283935

[5] Ensafi AA, Rezaloo F, Rezaei B. Electrochemical sensor based on porous silicon/silver nanocomposite for the determination of hydrogen peroxide. Sensors and Actuators B: Chemical. 2016;**231**:239-244. DOI: 10.1016/j.snb.2016.03.01

[6] Dimitrios PN, Georgia-Paraskevi N, editors. Nanotechnology and Biosensors. 1st ed. United Kingdom: Elsevier; 2018. p. 470. DOI: 10.1016/C2017-0-00358-0

[7] Georgia-Paraskevi N, Dimitrios PN, Christina S, Stephanos K, Spyridoula B, Nikolaos T. Nanobiosensors Based on Graphene Electrodes: Recent Trends and Future Applications. Elsevier; 2018. pp. 161-177. DOI: 10.1016/B978-0-08-101971-9.00007-7. Chapter 6

[8] Zhu Z, Garcia-Gancedo L, Flewitt AJ, Xie H, Moussy F, Milne WI. A critical review of glucose biosensors based on carbon nanomaterials: Carbon nanotubes and graphene. Sensors (Basel). 2012;**12**(5):5996-6022. DOI: 10.3390/s120505996

[9] Ponnusamy R, Chakraborty B, Rout CS. Pd-doped WO_3 nanostructures as potential glucose sensor with insight from electronic structure simulations. The Journal of Physical Chemistry. B. 2018;**122**(10):2737-2746. DOI: 10.1021/acs.jpcb.7b11642

[10] Jiang P, Wang Y, Zhao L, Ji C, Chen D, Nie L. Applications of gold nanoparticles in non-optical biosensors. Nanomaterials (Basel). 2018;**8**(12):977. DOI: 10.3390/nano8120977

[11] Arvind K, Amit S, Ashwani K, Ramesh C. Porous silicon filled with Pd/WO_3–ZnO composite thin film for enhanced H_2 gas-sensing performance. RSC Advances. 2017;**7**:39666-39675. DOI: 10.1039/C7RA05341J

[12] Hamdy ME, Del Carlo M, Hussein HA, Salah TA, El-Deeb AH, Emara MM, et al. Development of gold nanoparticles biosensor for ultrasensitive diagnosis of foot and mouth disease virus. Journal of Nanobiotechnology. 2018;**16**(1):1-12. DOI: 10.1186/s12951-018-0374-x

[13] Zhenga Q, Wua H, Wanga N, Yanc R, Mab Y, Guangb W, et al. Graphene-based biosensors for biomolecules detection. Current Nanoscience. 2014;**10**(5):627-637. DOI: 10.3390/mi11010060

[14] Kim HM, Park JH, Lee SK. Fiber optic sensor based on ZnO nanowires decorated by Au nanoparticles for improved plasmonic biosensor. Scientific Reports. 2019;**9**:15605. DOI: 10.1038/s41598-019-52056-1

[15] Feng C, Xu G, Liu H, Lv J, Zheng Z, Wu Y. Glucose biosensors based on Ag nanoparticles modified TiO_2 nanotube arrays. Journal of Solid State Electrochemistry. 2014;**18**:163-171. DOI: 10.1007/s10008-013-2257-2

[16] Izyumskaya N, Tahira A, Ibupoto ZH, Lewinski N, Avrutin V, Özgür Ü, et al. Electrochemical biosensors based on ZnO nanostructures. ECS Journal of Solid State Science and Technology. 2017;**6**(8):84-100. DOI: 10.1149/2.0291708jss

[17] Szunerits S, Boukherroub R. Graphene-based biosensors. Interface Focus. 2018;**8**:20160132. DOI: 10.1098/rsfs.2016.0132

[18] Bagyalakshmi S, Karthick AA. Study on enzymatic electrochemical glucose biosensors based on ZnO nanorods. International Journal of Scientific Research and Review. 2018;**7**(5):73-81. ISSN: 2279-543X

[19] Said Ragab AM, Hasan MA, Abdelzaher AM, Abdel-Raoof AM. Insights into the developments of nanocomposites for its processing and application as sensing materials. Journal of The Electrochemical Society. 2020;**167**(3):037549. DOI: 10.1149/1945-7111/ab697b

[20] Alabsi SS, Ahmed AY, Dennis JO, Md Khir MH, Algamili AS. A review of carbon nanotubes field effect-based biosensors. IEEE Access. 2020;**8**:69509-69521. DOI: 10.1109/ACCESS.2020.2987204

[21] Nag A, SC Mukhopadhyay SC, Kosel J. Wearable flexible sensors: A review. IEEE Sensors Journal. 2017;**17**(13):3949-3960. DOI: 10.1109/JSEN.2017.2705700

[22] Belorus A, Bespalova K, Spivak Y. Morphology and internal structure of porous silicon powders in dependence on the conditions of post-processing. In: Proceedings of the 2016 IEEE Conference of Russian Young Researchers in Electrical and Electronic Engineering (EIConRus); 2-3 February, 2016; Saint-Petersburg. Moscow: IEEE; 2016. pp. 23-28

[23] Pastukhov A, Belorus A, Bukina Y, Spivak Y, Moshnikov V. Influence of technology conditions on the surface energy of porous silicon using the method of contact angle. In: Proceedings of the 2017 IEEE Conference of Russian Young Researchers in Electrical and Electronic Engineering (EIConRus); 1-3 February, 2017; Saint-Petersburg. Moscow: IEEE; 2017. pp. 1183-1185

[24] Spivak Y, Belorus A, Somov P, Bespalova K, Moshnikov V, Tulenin S. Porous silicon nanoparticles for target drag delivery: Structure and morphology. Journal of Physics Conference Series. 2015;**643**:012022. DOI: 10.1088/1742-6596/643/1/012022

[25] Spivak Y, Belorus A, Moshnikov V, Bespalova K, Somov P, Panevin A, et al. Porous silicon as a nanomaterial for disperse transport systems of targeted drug delivery to the inner ear. The Russian Journal of Applied Physics. 2018;**63**:1352-1360. DOI: 10.1134/S1063784218090207

[26] Permiakov N, Matyushkin L, Belorus A, Koshevoi V. Investigation of a program-controlled process of impregnation of porous semiconductors with silver nanoparticles to create an electrical contact. In: Proceedings of the 2018 IEEE Conference of Russian Young Researchers in Electrical and Electronic Engineering (EIConRus); 29 January–1 February, 2018; Saint-Petersburg. Moscow: IEEE; 2018. pp. 539-543

[27] Belorus A, Pastukhov A, Vasyukov A, Moshnikov V. Formation

of ZnO nanoparticles using electrochemical deposition technique for applications in sensorics and biomedicine. In: 2019 IEEE Conference of Russian Young Researchers in Electrical and Electronic Engineering (EIConRus); 28-31 Jan. 2019; Saint-Petersburg. Moscow: IEEE; 2019. pp. 763-764

[28] Koshevoi V, Belorus A. Study of producing sensors based on porous layers of GaP: Te semiconductors with the use of electrodiffusion contacts. In: Proceedings of the 2017 IEEE Conference of Russian Young Researchers in Electrical and Electronic Engineering (EIConRus); 1-3 February, 2017; Saint-Petersburg. Moscow: IEEE; 2017. pp. 1406-1408

[29] Koshevoi V, Belorus A, Mikhailov I, Tarasov S, Solomonov A, Moshnikov V. Luminescent structures based on porous layers of gallium phosphide including embedded arrays of colloidal quantum dots of cadmium chalcogenides. In: Proceedings of the 2017 IEEE Conference of Russian Young Researchers in Electrical and Electronic Engineering (EIConRus); 1-3 February, 2017; Saint-Petersburg. Moscow: IEEE; 2017. pp. 1457-1459

[30] Pshchelko N, Vodkailo E, Tomaev V, Klimenkov B, Koshevoi V, Belorus A. Influence of electric field on adhesion and structure of conducting films on dielectric substances. Chemistry & Chemical Technology. 2017;**60**:101-104. DOI: 10.6060/tcct.2017608.5649

[31] Zolotarev V. Methods for the Study of Photonics Materials: Elements of Theory and Technology: A Training Manual. St. Petersburg: SPbSU ITMO; 2008. p. 275 (in Russian)

[32] Tolstoy V, Chernyshova I, Skryshevsky V. Handbook of Infrared Spectroscopy of Ultrathin Films. John Wiley & Sons Inc.; 2003. p. 710. DOI: 10.1002/047123432X

[33] Zanoni R, Righini G, Mattogno G, Schirone L, Sotgiu G, Rallod F. X-ray photoelectron spectroscopy characterization of stain-etched luminescent porous silicon films. Journal of Luminescence. 1998;**80**:159-162. DOI: 10.1016/S0022-2313(98)00088-X

[34] Leisenberger F, Duschek R, Czaputa R, Netzer FP, Beamson G, Matthew JAD. A high-resolution XPS study of a complex insulator. Applied Surface Science. 1997;**108**:273-281. DOI: 10.1016/S0169-4332(96)00595-8

[35] Jordan P, Todd G. Simple optical method to determine the porosity of porous silicon films. Thin Solid Films. 2012;**520**:2526-2531. DOI: 10.1016/j.tsf.2011.10.146

[36] Nossarzewska-Orlowska E, Brzozowski A. Photoluminescence properties of porous silicon prepared by electrochemical etching of Si epitaxial layer. In: Proceedings of the XXII International School of Semiconducting Compounds; Jaszowiec. Jaszowiec: Institute of Electronic Materials Technology; 1993. pp. 713-716

Section 2

Composite Materials with Other Kind of Matrix

Chapter 8

Investigation of Shielding Effectiveness of M-Type Ba-Co-Ti Hexagonal Ferrite and Composite Materials in Microwave X-Band Systems

Charanjeet Singh, S. Bindra Narang and Ihab A. Abdel-Latif

Abstract

Ferrites are a wide class of materials that are still a very rich field of scientific interest and under the scope of recent research. The polycrystalline Co^{2+}-Ti^{4+} substituted Ba hexagonal ferrite has been synthesized by the standard ceramic method. The vector network analyzer has been incorporated to measure different microwave parameters at X-band (8.2–12.4 GHz) frequencies. The microwave shielding effectiveness is evaluated by S-parameters for near field and AC conductivity as well as skin depth for far field. The doping of Co^{2+} and Ti^{4+} ions causes absorption in composite x = 0.5 to exhibit good shielding effectiveness and it exhibits large 20-dB bandwidth of 4.70 GHz in the near field and 3.60 GHz for far field respectively. The AC conductivity increases with frequency in composites x = 0.1, 0.3, and 0.5 and skin depth decreases with frequency in all composites. The shielding effectiveness, AC conductivity, and skin depth are correlated to each other.

Keywords: ferrites, hexaferrite, microwave shielding, AC conductivity

1. Introduction

Ferrites are a wide class of materials containing iron. These materials are formed in different crystalline symmetries. A simple form of ferrites is the spinel AB_2O_4 of cubic structure [1–9]. The orthoferrites ABO_3 are another important form with an orthorhombic perovskite crystal system [10–19]. The third class of ferrites are garnets of form $A_3B_5O_{12}$ [20–29]. The fourth class, termed as hexaferrites, may be divided into five main groups: M-type ($AB_{12}O_{19}$), W-type ($AMe_2B_{16}O_{27}$), X-type $A_2Me_2B_{28}O_{46}$, Y-type $A_2Me_2B_{12}O_{22}$, and Z-type or $A_3Me_2B_{24}O_{41}$ [30–49]. The preparation of these materials and their characterization are very rich topics because of the wide range of applications and the cheap materials obtained. Ferrites are a very interesting class of materials whose wide range of applications are related to electromagnetic interference suppression as well as their use in radar absorbing material (RAM) coatings [50]. From this point of view, great scientific interests are devoted to use these materials as RAM devices [51–54]. In this work the intensive highlights is devoted to the microwave applications and which class is the best candidate for this application.

IntechOpen

The tremendous rise in speed of electronic devices and widespread incorporation of information technology for various technological applications have pumped up electromagnetic pollution to dangerous levels. The high-speed electronic gadgets emit spurious wireless signals rendering the electromagnetic disturbance/interference (EMI) to the electrical and/or electronic circuits in the vicinity.

A microwave absorber reduces unwanted radiation emitted from high-speed electronic devices such as radar, oscillators, and supercomputers. The ferrimagnetic materials ferrites have the potential ability to reduce electromagnetic interference (EMI) in contrast with conventional dielectrics owing to their magneto/dielectric properties [55–60]. Electronic devices constitute integrated circuits (ICs) wherein numerous components are embedded and such components are encapsulated with ferrite films to mitigate EMI. The frequency range of application of extensively used spinel ferrites is limited by Snoek's limit and they are not effective at GHz range. M-type hexagonal ferrites are tailored for EMI diminution in the higher end of microwave region, that is, X-band, Ku-band, K-band, etc. [61–65]: these ferrites allow to tune in the frequency region through doping accompanied by anisotropy field. Both the electric and magnetic properties define the capabilities of these materials to store energy and are described by analyzing the real parts of complex permittivity (ε') and permeability (μ'), respectively. On the other side, imaginary parts (ε'', μ'') are very important parameters that describe the loss of electric and magnetic energy.

Different researches have been devoted to electromagnetic interference (EMI) shielding effectiveness (SE) and EMI shielding mechanisms [66–68] of high structure carbon black (HS-CB)/polypropylene (PP) composites and multiwalled carbon nanotubes-polymethyl methacrylate (MWCNT-PMMA) in the X-band frequency range. They studied different thickness of composite plates electrical conductivities. Their results showed that the absorption loss contribution to the overall attenuation is more than the contribution of the reflection loss for HS-CB/PP composites. Moreover, EMI SE up to 40 dB in the frequency range 8.2–12.4 GHz (X-band) was achieved in Ref. [69] by stacking seven layers of 0.3-mm-thick MWCNT-PMMA composite films compared with 30 dB achieved by stacking two layers of 1.1-mm-thick MWCNT-PMMA bulk composite.

Recently, graphene composites have been found to be one of the most promising candidates for high-performance porous microwave absorbers in ref. [70] because of their 3D conductive network and multiple scattering. A qualified frequency bandwidth (reflection loss <-10 dB) reaches 5.28 GHz covering almost the entire Ku band at 2 mm thickness. These results might open the door for a new design of lightweight coating absorber. This may allow us to say that the performance of microwave devices is mainly based on the properties of the used materials. Knowledge of the frequency dependence of such material is a prerequisite to select suitable materials for various microwave applications and vice versa [71–73]. Novel nanocomposite systems are prepared for microwave applications such as *para*-toluene sulfonic acid (p-TSA)-doped polyaniline (PANI)-graphene nanoplatelet (GRNP) composite films. The addition of GRNPs in the PANI matrix allows to improve the conductivity and dielectric properties of the composites due to the formation of 3D conducting networks. Shielding effectiveness of the PANI-GRNP composite films doped with p-TSA was examined by using S-parameters obtained from vector network analyzer in the X-band microwave frequencies. The efficiency of shielding for these composites depends on GRNP's content in the PANI matrix [74]. Electrical and mechanical properties of carboxylic (–COOH) functionalized multiwall carbon nanotube (MWNTs)/epoxy composites at low wt.% (0.5, 0.75, and 1 wt.%) are studied in Ref. [75]. Microwave shielding effectiveness (SE) for X-band (8–12 GHz) and the flexural properties showed that the total SE of the nanocomposites was increased with the positive gradient of MWNT contents. Great efforts have been made to improve the requirements for microwave applications in X-band and new materials are being tested [70, 76–78]. Promising results were found and the search for new materials continues.

In the present chapter, we have explored EMI shielding effectiveness characteristics of M-type Ba-Co-Ti hexagonal ferrites.

2. Experimental details

The M-type $BaCo_xTi_xFe_{(12-2x)}O_{19}$ hexaferrites, with x = 0.1, 0.3, 0.5, and 0.7, were prepared by ceramic method. The powder chemicals were mixed thoroughly, ground, and sintered in an electric furnace at 900°C for 7 h. The pellets were made of the powder with the hydraulic press at uniform pressure of 75 kN/m^2 and final sintering was done at 1100°C for 9 h. The crystal structure was measured using Bruker D8Diffractometer of Cu X-ray radiation.

The microwave properties have been studied by the vector network analyzer, Agilent model N5225A. Before performing the measurements, permittivity and permeability of air were measured with an analyzer for calibration purposes. The DC resistivity (ρ_{dc}) was investigated using Keithley Electrometer, model 6514. The selected thickness of composites for optimized characteristics are x = 0.1–3.3 mm, x = 0.3–3.8 mm, x = 0.5–3.4 mm, and x = 0.7–3.2 mm.

3. Results and discussion

Figure 1 shows patterns obtained from X-ray diffraction of $BaCo_xTi_xFe_{(12-2x)}O_{19}$ hexaferrite composites. The observed XRD peaks confirm M-type phase of hexaferrite with space group P6$_3$/*mmc*. The change of intensity in the peaks shows that the substituted Co^{2+} and Ti^{4+} ions have occupied crystallographic sites.

3.1 Shielding in near field

The shielding effectiveness (SE) is accompanied by reflection or absorption of unwanted microwave signal (EMI) and can be represented as $SE = SE_R + SE_A$ with SE_R due to reflection and SE_A as absorption. When the microwave signal passes

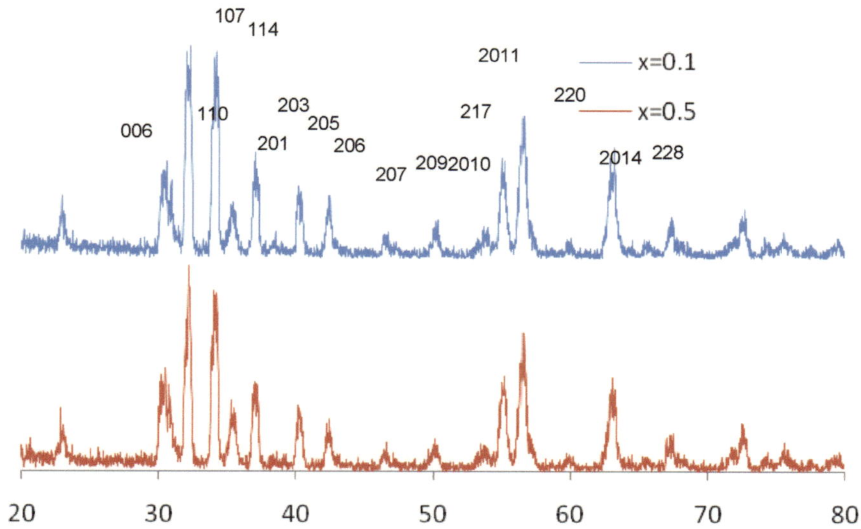

Figure 1.
X- ray diffraction pattern of $BaCo_xTi_xFe_{(12-2x)}O_{19}$ ferrite.

through the material, part of the signal is reflected and remaining transmitted or absorbed. The reflected power (P_r) and transmitted power (P_t) are derived from measured S-parameters: $P_r = |S_{11}|^2$ and $P_t = |S_{22}|^2$, SE_A and SE_R can be calculated as:

$$SE_A = -10\log\left[P_t/(1-P_r)\right] \quad (1)$$

$$SE_R = -10\log(1-P_r) \quad (2)$$

Figure 2 shows plots of EMI shielding effectiveness (SE_A) versus frequency for doping of Co^{2+} and Ti^{4+} ions. Composites x = 0.5 and 0.7 exhibit highest (38.9 dB) and lowest (7.9 dB) values at 10.26 and 12.03 GHz respectively and these composites stay at maximum and minimum values in the frequency regime.

All composites exhibit nonlinear decrease in SE_A with frequency and composites x = 0.1, 0.3, and 0.5 show more dispersion in SE_A with frequency: x = 0.1, 0.3, and 0.7 displaying maxima at 9.27 GHz and x = 0.5 at 10.26 GHz. All composites stay at $SE_A > 10$ dB or 90% absorption, encompassing the entire frequency region.

Figure 3 depicts the response of shielding effectiveness (SE_R) of $BaCo_xTi_xFe_{(12-2x)}O_{19}$ ferrite versus frequency for doping of Co^{2+} and Ti^{4+} ions. All composites exhibit: (i) minimum SE_R in comparison to SE_A encompassing the entire frequency region, (ii) nearly the same trend of SE_R in the investigated frequency regime, and (iii) maxima in the mid-frequency region. The shielding effectiveness (SE_R) due to reflection is very small and SE_R owe excursion between 0.2 and 2.5 dB. The low SE_R implies no composite can act as a microwave reflector shield. The composite x = 0.5 has largest SE_R = 2.48 dB at 10.39 GHz.

3.2 Shielding in far field

Shielding effectiveness for far field can be evaluated by classical electromagnetic field theory with the following relation [79]:

$$SE(dB) = 10\log(\sigma_{ac}/16\omega\varepsilon_o) + 20(d/\delta)\log e \quad (3)$$

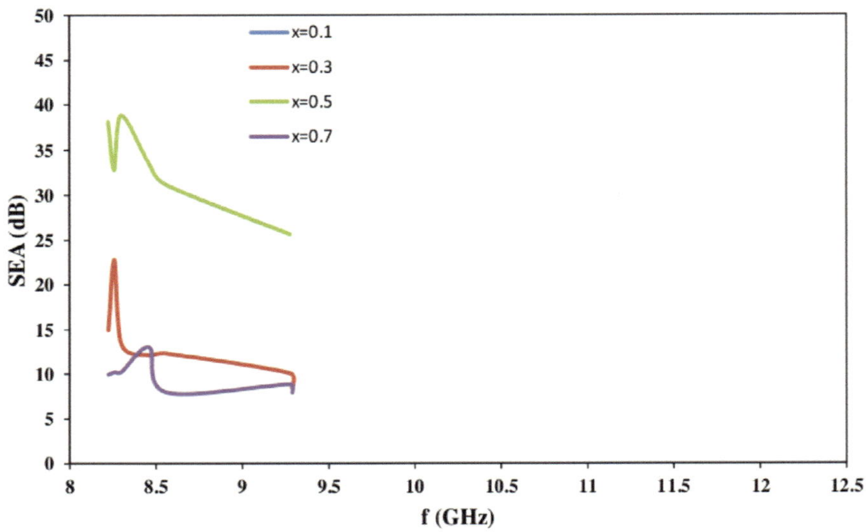

Figure 2.
Variation of shield effectiveness due to absorption (SE_A) with frequency for $BaCo_xTi_xFe_{(12-2x)}O_{19}$ ferrite (x = 0.1, 0.3, 0.5, 0.7).

where σ_{ac} is the AC conductivity, ω is the angular frequency, ε_0 is the absolute permittivity, d is the thickness of the shield, δ is the skin depth, and μ_r is the relative permeability.

Furthermore, $\sigma_{ac} = \omega\varepsilon_0\varepsilon''$ and $\delta = (2/\mu_o\omega\sigma_{ac})^{1/2}$, where μ_o and ε'' are dielectric loss and absolute permeability respectively. The first term, $10\log(\sigma_{ac}/16\omega\varepsilon_0)$, in Eq. (3) is the shielding effectiveness due to reflection and second term, $20(d/\delta)\log e$, relates to the absorption of the microwave signal. The second term is effective at high frequencies and Eq. (3) can be rewritten as:

$$SE_A = 20d(\mu_o\omega\sigma_{ac}/2)^{1/2}\log e \quad (4)$$

Figure 4 depicts the graph of AC conductivity (σ_{ac}) as a function of frequency for doping of Co^{2+} and Ti^{4+}. It increases with doping from x = 0.1, 0.3 and x = 0.5 followed by prevalent fall in x = 0.7: composite x = 0.5 observes more dispersion with frequency and large value of σ_{ac} in comparison to other composites. The rise

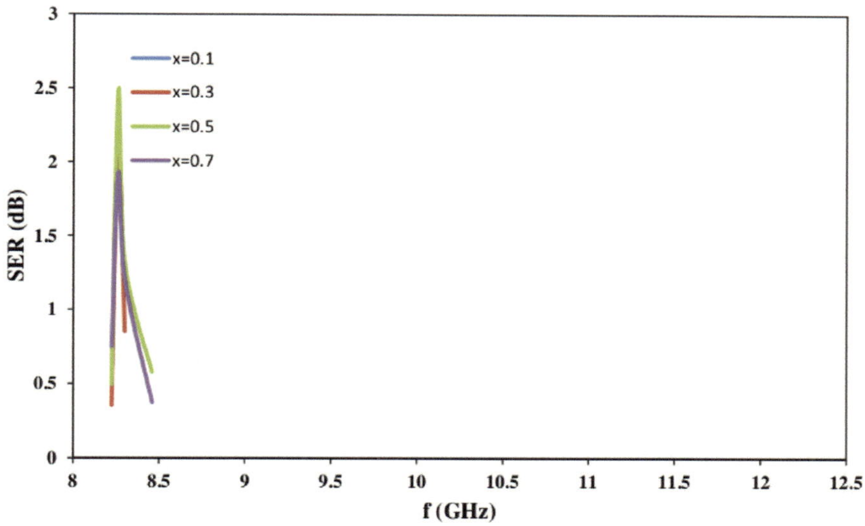

Figure 3.
Variation of shield effectiveness due to reflection (SE_R) with frequency for $BaCo_xTi_xFe_{(12-2x)}O_{19}$ ferrite (x = 0.1, 0.3, 0.5, 0.7).

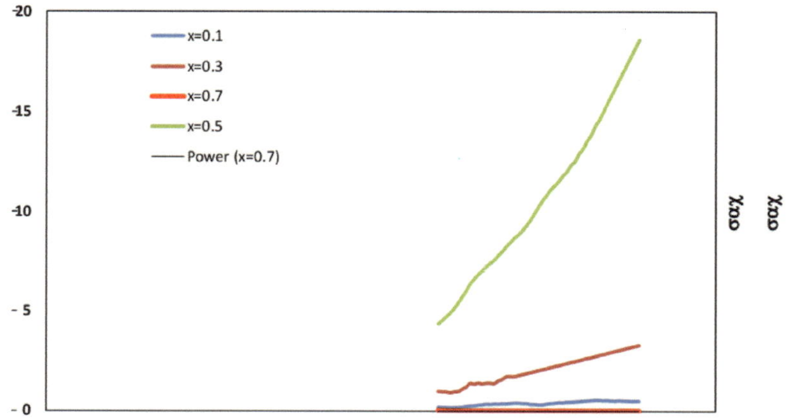

Figure 4.
Plots of AC conductivity versus frequency for $BaCo_xTi_xFe_{(12-2x)}O_{19}$ ferrite (x = 0.1, 0.3, 0.5, 0.7).

in σ_{ac} is seen with frequency in composite x = 0.1, 0.3, and 0.5; however, it remains nearly independent of frequency in x = 0.7. This increase in σ_{ac} is ascribed to Koops-Wagner model, which explains ferrite comprising of heterogeneous structure [80]: ferrites owe layers of good conducting grains, effective at high frequencies, are separated by poor conducting grain boundaries that are effective at low frequencies.

The composites x = 0.1, 0.3, 0.5, and 0.7 have DC resistivity (ρ_{dc}) of 693.6 MΩ cm, 2.8 kΩ cm, 0.5 kΩ cm, and 33.8 MΩ cm, respectively. The composite x = 0.1 has the highest resistivity but still a large σ_{ac} attributed to the presence of more strength of Fe^{3+}: electron hopping between Fe^{3+}–Fe^{2+} ions is responsible for conduction in ferrites [81]. Among all composites, composite x = 0.5 (i) owe maximum σ_{ac} besides with diminution in the number of Fe^{3+} ions and (ii) has the lowest DC resistivity. The competition between these factors altogether increases σ_{ac} in this

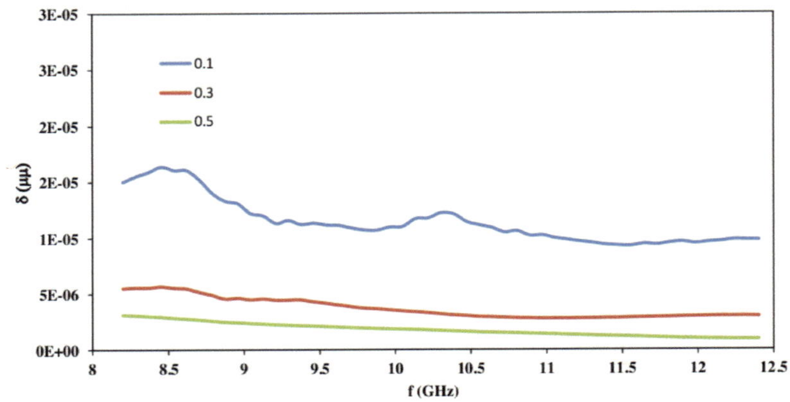

Figure 5.
Change in skin depth (δ) with frequency for $BaCo_xTi_xFe_{(12-2x)}O_{19}$ ferrite (x = 0.1, 0.3, 0.5, 0.7) with frequency in X-band.

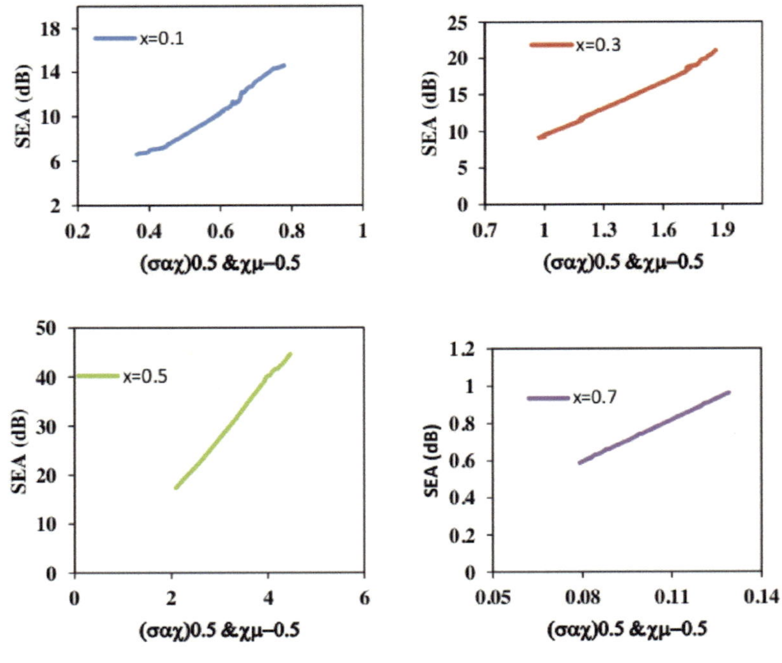

Figure 6.
Plots of SE_A versus $(\sigma_{ac})^{0.5}(S/m)^{0.5}$ for $BaCo_xTi_xFe_{(12-2x)}O_{19}$ ferrite (x = 0.1, 0.3, 0.5, 0.7).

x	Near field				Far field			
	Freq. band (GHz)	10 dB bandwidth (GHz)	Freq. band (GHz)	20 dB bandwidth (GHz)	Freq. band (GHz)	10 dB bandwidth (GHz)	Freq. band (GHz)	20 dB bandwidth (GHz)
0.1	8.26–8.59	0.33	8.59–9.80	1.54	9.20–12.40	3.20	–	–
	9.80–12.03	2.23	–	–	–	–	–	–
0.3	8.26–8.80	0.54	8.80–9.69	0.89	8.70–12.40	3.70	–	–
	9.69–12.03	2.34	–	–	–	–	–	–
0.5	–	–	8.30–11.90	3.60	8.20–8.70	0.50	8.70–12.40	4.70
0.7	8.20–10.32	2.12	–	–	–	–	–	–

Table 1.
Microwave shielding effectiveness (SE_A) for 10- and 20-dB bandwidth (BW) in near and far field in $BaCo_xTi_xFe_{(12-2x)}O_{19}$ (x = 0.1, 0.3, 0.5, 0.7).

composite. Similarly, steep fall of σ_{ac} in x = 0.7 is associated with the least number of Fe^{3+} ions available for electron hopping and large DC resistivity.

The dependence of skin depth (δ) on frequency for a different level of substitution is shown in **Figure 5**. The decrement trend in δ is observed with frequency, and x = 0.7 and 0.5 exhibit large and small δ respectively among the composites in the frequency regime. The large conduction loss, as shown in σ_{ac} (**Figure 4**), causes minimum δ, which attenuates the propagating microwave signal in the composite and vice versa; thus further penetration of signal is not possible inside the thickness of composite: the signal is attenuated more in x = 0.5 due to highest σ_{ac} depicted in **Figure 4**, thereby causing lowest δ.

The dependence of shielding effectiveness (SE_A) on AC conductivity ($\sigma_{ac}^{0.5}$) for different levels of doping is shown in **Figure 6**: it increases with doping from x = 0.1 to x = 0.5 and steep decrement is seen thereafter in x = 0.7. All composites display a monotonic trend of increase in SE_A with $\sigma_{ac}^{0.5}$ and x = 0.5 owe maximum value while x = 0.7 stay at lowest one.

Table 1 shows bandwidth (10 dB and 20 dB) of SE_A for both near and far field versus doping: 10 and 20 dB means 90% and 99% absorption respectively. For near field, x = 0.1, 0.3, and 0.7 exhibit 10-dB bandwidth of 2.23, 2.34, and 2.12 GHz respectively whereas 20-dB bandwidth of 1.54, 0.89, and 3.60 GHz is observed in x = 0.1, 0.3 and 0.5 respectively. For far field, x = 0.1, 0.3, and 0.5 show 10 dB-bandwidth of 3.20, 3.70, and 0.50 GHz respectively, and 20-dB bandwidth of 4.70 GHz is seen in x = 0.5 only.

4. Conclusions

For near and far field, microwave shielding effectiveness in $BaCo_xTi_xFe_{(12-2x)}O_{19}$ ferrite is governed by absorption and doping of Co^{2+} and Ti^{4+} ion increases SE_A from x = 0.1, 0.3, and 0.5. Composite x = 0.5 owes the highest SE_A of 38.9 dB at 10.26 GHz and 3.4 mm thickness; $\sigma_{ac}^{0.5}$, ρ_{dc} and δ are the contributing factors and same composite carries with highest SE_A of 44.6 dB at $\sigma_{ac}^{0.5}$ of 4.5 $(Ohm.cm)^{-0.5}$ for far field; s-parameter is the deciding factor. Furthermore, SE_A increases monotonically with frequency and it can be tuned by varying intrinsic and extrinsic parameters. Composite x = 0.5 has far field and near field wideband of 4.70 and 3.60 GHz respectively for 20 dB SE_A. The studied composites have the potential for practical absorber applications. The applications of these composite materials or other composite materials are very an important subject and more research is needed to find the optimum properties and optimum materials for X-band microwave applications.

Acknowledgements

The author *IA Abdel-Latif*, is thankful to the Deanship of Scientific Research in Najran University for their financial support NU/ESCI/16/063 in the frame of the local scientific research program support.

Author details

Charanjeet Singh[1*], S. Bindra Narang[2] and Ihab A. Abdel-Latif[3,4,5*]

1 Department of Electronics and Communication Engineering, Lovely Professional University, Phagwara, Punjab, India

2 Department of Electronics Technology, Guru Nanak Dev University, Amritsar, Punjab, India

3 Physics Department, College of Science and Arts, Najran University, Najran, Kingdom of Saudi Arabia

4 Advanced Materials and Nano-Research Centre, Najran University, Najran, Saudi Arabia

5 Reactor Physics Department, NRC, Atomic Energy Authority, Abou Zabaal, Cairo, Egypt

*Address all correspondence to: rcharanjeet@gmail.com; charanjeet2003@rediffmail.com and ihab_abdellatif@yahoo.co.uk

References

[1] Gismelseed AM, Khalaf KAM, Elzain ME, Widatallah HM, Al-Rawas AD, Yousif AA. The structural and magnetic behavior of the $MgFe_{2-x}CrxO_4$ spinel ferrite. Hyperfine Interactions. 2012. DOI: 10.1007/s10751-011-0529-8

[2] Roumaih K, Manapov RA, Sadykov EK, Pyataev AV. Mossbauer studies of $Cu_{1-x}Ni_xFeMnO_4$ spinel ferrites. Journal of Magnetism and Magnetic Materials. 2005;**288**:267

[3] Abdel Latif IA. Fabrication of nano-size nickel ferrites for gas sensors applications. Journal of Physics. 2012;**1**(2):50-53

[4] Khalaf KA, Al-Rawas A, Gismelseed A, Al-Ruqeishi M, Al-Ani S, Al-Jubouri A, et al. Effects of Zn substitution on structure factors, Debye-Waller factors and related structural properties of the $Mg_{1-x}Zn_xFeNiO_4$ spinels. Advances in Materials. 2019;**8**(2):70-93. DOI: 10.11648/j.am.20190802.15

[5] Maensiri S, Sangmanee M, Wiengmoon A. Magnesium ferrite ($MgFe_2O_4$) nanostructures fabricated by electrospinning. Nanoscale Research Letters. 2009;**4**:221

[6] Fayek MK, Ata-Allah SS. ^{57}Fe Mossbauer and electrical studies of the $(NiO)–(Cr_2O_3)_x–(Fe_2O_3)_{2-x}$ system. Physica Status Solidi A: Applications and Material Science. 2003;**198**(457-464)

[7] Tatarchuk T, Bououdina M, Judith VJ, John Kennedy L. Spinel ferrite nanoparticles: Synthesis, crystal structure, properties, and perspective applications. In: Fesenko O, Yatsenko L, editors. Nanophysics, Nanomaterials, Interface Studies, and Applications. NANO 2016. Springer Proceedings in Physics, vol 195. Springer, Cham. 2017

[8] Zaki HM et al. Synthesis and characterization of nanocrystalline $MgAl_xFe_{2-x}O_4$ ferrites. Journal of Materials Research. 2012;**27**(21):2798

[9] Al-Maashani M, Gismelseed AM, Khalaf KAM, Yousif AA, Al-Rawas AD, Widatallah HM, et al. Structural and Mossbauer study of nanoparticles $CoFe_2O_4$ prepared by sol-gel auto-combustion and subsequent sintering. Hyperfine Interactions. 2018;**239**:15. DOI: 10.1007/s10751-018-1491-5

[10] Yousif AA et al. Study on Mossbauer and magnetic properties of strontium doped neodymium ferrimanganites perovskite-like structure. AIP Conference Proceedings. 2011;**1370**:103

[11] Abdel-Latif IA, Saleh SA. Effect of iron doping on the physical properties of europium Manganites. Journal of Alloys and Compounds. 2012;**530**:116

[12] Bashkirov S et al. Crystal structure, electric and magnetic properties of ferrimanganite $NdFe_xMn_{1-x}O_3$. Izv. RAS, Physical Series. 2003;**67**:1072

[13] Abdel-Latif IA et al. The influence of tilt angle on the CMR in $Sm_{0.6}Sr_{0.4}MnO_3$. Journal of Alloys and Compounds. 2008;**452**:245

[14] Abdel-Latif IA et al. Magnetocaloric effect, electric, and dielectric properties of $Nd_{0.6}Sr_{0.4}Mn_xCo_{1-x}O_3$ composites. Journal of Magnetism and Magnetic Materials. 2018;**457**:126

[15] Bouziane KA et al. Electronic and magnetic properties of $SmFe_{1-x}Mn_xO_3$ orthoferrites (x = 0.1, 0.2 and 0.3). Journal of Applied Physics. 2005;**97**(10A):504

[16] Abdel-Latif IA. Study on the effect of particle size of strontium-ytterbium manganites on some physical properties. AIP Conference Proceedings. 2011;**1370**:108

[17] Parfenov VV, Bashkirov SS, Abdel-Latif IA, Marasinskaya AV. Russian Physics Journal. 2003;**46**:979-983

[18] Ahmed Farag IS et al. Preparation and structural characterization of $Eu_{0.65}Sr_{0.35}Mn_{1-x}Fe_xO_3$. Egyptian Journal of Solids. 2007;**30**(1):149

[19] Abdel-Latif IA. Study on structure, electrical and dielectric properties of $Eu_{0.65}Sr_{0.35}Fe_{0.3}Mn_{0.7}O_3$. IOP Conference Series: Materials Science and Engineering. 2016;**146**:012003

[20] Ding J, Yang H, Miao WF, Mccormick PG, Street R. Journal of Alloys and Compounds. 1995;**221**:70-73

[21] Niaz Akhtar M et al. $Y_3Fe_5O_{12}$ nano particulate garnet ferrites: Comprehensive study on the synthesis and characterization fabricated by various routes. Journal of Magnetism and Magnetic Materials. 2014;**368**:393-400

[22] Yu H, Zeng L, Lu C, Zhang W, Xu G. Materials Characterization. 2011;**62**:378-381

[23] Rastogi AC, Moorthy VN. Materials Science and Engineering. 2002;**B95**:131-136

[24] Sanchez RD, Rivas J, Vaqueiro P, Quintela MAL, Caeiro D. Journal of Magnetism and Magnetic Materials. 2002;**247**:92-98

[25] Liu CP, Li MW, Cui Z, Huang JR, Tian YL, Lin T, et al. Journal of Materials Science. 2007;**42**:6133-6138

[26] Abbas Z, Al-habashi RM, Khalid K, Maarof M. European Journal of Scientific Research. 2009;**36**:154-160

[27] Rajendran M, Deka S, Joy PA, Bhattacharya AK. Journal of Magnetism and Magnetic Materials. 2006;**301**:212-219

[28] Verma S, Pradhan SD, Pasricha R, Sainkar SR, Joy PA. Journal of the American Ceramic Society. 2005;**88**:2597-2259

[29] Aen F, Ahmad M, Rana MU. Current Applied Physics. 2013;**13**:41-46

[30] Singh J et al. Elucidation of phase evolution, microstructural, Mossbauer and magnetic properties of CO_2-Al_3 doped M-type Ba-Sr hexaferrites synthesized by a ceramic method. Journal of Alloys and Compounds. 2017;**695**:1112-1121

[31] Singh C, Narang SB, Hudiara IS, Bai Y, Marina K. Hysteresis analysis of Co–Ti substituted M-type Ba–Sr hexagonal ferrite. Materials Letters. 2009;**63**:1921-1924

[32] Sharbati A, Choopani S, Azar A-M, Senna M. Structure and electromagnetic behavior of nanocrystalline $SrMg_xZr_xFe_{12-2x}O_{19}$ in the 8-12 GHz frequency range. Journal of Solid State Communications. 2010;**150**:2218-2222

[33] Reimann T, Schmidt T, Töpfer J. Phase stability and magnetic properties of $SrFe_{18}O_{27}$ W-type hexagonal ferrite. Journal of the American Ceramic Society. 2019. DOI: 10.1111/jace.16726

[34] Arjunwadkar PR, Salunkhe MY, Dudhe CM. Structural, electrical, and magnetic study of $SrNi^{2+}(Li^{1+}Fe^{3+})_{0.5}Fe_{16}O_{27}$ ferrite. Journal of Solid State Physics. 2013. DOI: 10.1155/2013/471472

[35] MukhtarA, GrossingerR, KriegischM, Kubel F, Rana MU. Characterization of Sr-substituted W-type hexagonal ferrites synthesized by sol–gel autocombustion method. Journal of Magnetism and Magnetic Materials. 2013;**332**:137-145

[36] Meshram MR, Agrawal NK, Sinha B, Misra PS. Journal of Magnetism and Magnetic Materials. 2004;**271**:207

[37] Goldman A. Modern Ferrite Technology. 2nd ed. Springer Publication; 2006. pp. 83

[38] Amer MA, Hemeda OM. Hyperfine Interactions. 1995;**96**:99

[39] Hosaka N et al. Crystal structure and magnetic properties of X-type hexagonal ferrite $Ba_2Ni_2Fe_{28}O_{46}$. Journal of the Japan Society of Powder and Powder Metallurgy. 2010;**57**:41-45

[40] Gu BX. Magnetic properties of *X*-type $Ba_2Me_2Fe_{28}O_{46}$ (Me = Fe, Co, and Mn) hexagonal ferrites. Journal of Applied Physics. 1992;**71**:5103. DOI: 10.1063/1.350613

[41] Gu BX. Magnetic properties of $Ba_2Me_2Fe_{28}O_{46}$ (Me_2-*X*, Me = Ni, Cu, Mg, and Zn) hexaferrites. Journal of Applied Physics. 1991;**70**:372. DOI: 10.1063/1.350284

[42] Ben-Xi G, Huai-Xian LU, You-Wei DU. Magnetic properties and Mössbauer spectra of X type hexagonal ferrites. Journal of Magnetism and Magnetic Materials. 1983;**31-34**(Part 2):803-804

[43] Reddy MB, Reddy PV. Low-frequency dielectric behaviour of mixed Li-Ti ferrites. Journal of Physics D: Applied Physics. 1991;**24**:975

[44] Bayrakdar H. Fabrication, magnetic and microwave absorbing properties of $Ba_2Co_2Cr_2Fe_{12}O_{22}$ hexagonal ferrites. Journal of Alloys and Compounds. 2016;**675**:185-188

[45] Bai Y, Zhou J, Gui Z, Yue Z, Li L. Complex Y-type hexagonal ferrites: An ideal material for high-frequency chip magnetic components. Journal of Magnetism and Magnetic Materials. 2003;**264**(1):44-49

[46] Lee SG, Kwon SJ. Saturation magnetizations and Curie temperatures of Co-Zn Y-type ferrites. Journal of Magnetism and Magnetic Materials. 1996;**153**:279-284

[47] Smit J, Wijn HPJ. Ferrites. Eindhoven, The Netherlands: Philips Technical Library; 1965. p. 177

[48] Sugimoto M. In: Wohlfarth EP, editor. Ferromagnetic Materials. Vol. 3. Amsterdam: North-Holland; 1982. p. 393

[49] Lubitz P, Rachford FJ. Z type Ba hexagonal ferrites with tailored microwave properties. Journal of Applied Physics. 2002;**91**:7613. DOI: 10.1063/1.1453932

[50] Nakamura T, Hankui E. Control of high-frequency permeability in polycrystalline (Ba,Co)-Z-type hexagonal ferrite. Journal of Magnetism and Magnetic Materials. 2003;**257**(2-3):158-164

[51] Narang SB, Kaur P, Bahel S, Singh C. Microwave characterization of Co–Ti substituted barium hexagonal ferrites in X-band. Journal of Magnetism and Magnetic Materials. 2016;**405**:17-21

[52] Zhang BS, Feng Y, Xiong J, Yang Y, Lu HX. Microwave-absorbing properties of de-aggregated flake-shaped carbonyl-iron particle composites at 2-18GHz. IEEE Transactions on Magnetics. 2006;**42**:1778-1781

[53] Singh P, Babbar VK, Razdan A, Srivastava SL, Puri RK. Complex permeability and permittivity, and microwave absorption studies of $Ca(CoTi)_xFe_{12-2x}O_{19}$ hexaferrite composites in X-band microwave frequencies. Materials Science and Engineering. 1999;**B67**:132-138

[54] Sugimoto S, Haga K, Kagotani T, Inomata K. Microwave absorption properties of BaM-type ferrite prepared by a modified co precipitation method. Journal of Magnetism and Magnetic Materials. 2005;**290**:1188-1191

[55] Singh C, Narang SB, Hudiara IS, Sudheendran K, James Raju KC. Complex permittivity and complex permeability of Sr ions substituted Ba ferrite at X-band. Journal of Magnetism and Magnetic Materials. 2008;**3**(20):1657-1665

[56] Lima UR, Nasar MC, Rezende MC, Araugo JH. Journal of Magnetism and Magnetic Materials. 2008;**320**:1666

[57] Aphesteguy JC, Pamiani A, Digiovanni D, Jacobo SE. Physica B. 2009;**404**:2713

[58] Huang X, Zhang J, Lai M, Sang T. Journal of Alloys and Compounds. 2015;**627**:367

[59] Koledintseva MY, Mikhailovsky LK, Kitaytsev AA. IEEE Transactions on Electromagnetic Compatibility. 2000;**2**:773

[60] Meng P, Xiong K, Ju K, Li S, Xu G. Journal of Magnetism and Magnetic Materials. 2015;**385**:407

[61] Liu J, Zhang J, Zhang P, Wang S, Lu C, Li Y, et al. Materials Letters. 2015;**158**:53

[62] Meng P, Xiong K, Wang L, Li S, Cheng Y, Xu G. Journal of Alloys and Compounds. 2015;**628**:75

[63] Singh C, Bindra Narang S, Vikramjit Singh, Kotnala RK. IEEE 15th International Symposium on Antenna Technology and Applied Electromagnetics (ANTEM), Toulouse; 2012

[64] Singh C, Bindra Narang S, Hudiara IS. IEEE XXX General Assembly and Scientific Symposium of the International Union of Radio Science (URSI), Istanbul; 2011

[65] Singh C, Bindra Narang S, Hudiara IS, Koledintseva MY, Kitaitsev AA. Asia-Pacific Radio Science Conference (AP-RASC'10) Toyama. Paper E4-3; 2010

[66] Al-Saleh MH, Sundararaj U. X-band EMI shielding mechanisms and shielding effectiveness of high structure carbon black/polypropylene composites. Journal of Physics D: Applied Physics. 2012;**46**(3):035304

[67] Cui R-B, Zhang C, Zhang J-Y, Xue W, Hou Z-L. Highly dispersive GO-based supramolecular absorber: Chemical-reduction optimization for impedance matching. Journal of Alloys and Compounds. 2020;**155122**

[68] Raveendran A, Sebastian MT, Raman S. Applications of microwave materials: A review. Journal of Electronic Materials. 2019;**48**(5):2601-2634

[69] Pande S, Singh BP, Mathur RB, Dhami TL, Saini P, Dhawan SK. Improved electromagnetic interference shielding properties of MWCNT? PMMA composites using layered structures. Nanoscale Research Letters. 2009;**4**(4):327-334

[70] Kashi S, Hadigheh SA, Varley R. Microwave attenuation of graphene modified thermoplastic poly (butylene adipate-co-terephthalate) nanocomposites. Polymers. 2018;**10**(6):582

[71] Abdel-Latif IA. Crystal structure and electrical transport of nano-crystalline strontium-doped neodymium ortho-ferrites. Journal of Nanoparticle Research. 2020;**22**:60

[72] Saleh S, Abdel-Latif I, Hakeem AA, et al. Structural and frequency-dependent dielectric properties of $(SnO_2)_{1-x}$ $(Fe_2O_3)_x$. Journal of Nanoparticle Research. 2020;**22**:44. DOI: 10.1007/s11051-020-4763-3

[73] Abdel-Latif IA. The particle size effect of $Yb_{0.8}R_{0.2}MnO_3$ (R is

Sm, Nd, and Eu) on some physical properties. Journal of Nanoparticle Research. 2020;**22**:45. DOI: 10.1007/s11051-020-4759-z

[74] Khasim S. Polyaniline-graphene nanoplatelet composite films with improved conductivity for high performance X-band microwave shielding applications. Results in Physics. 2019;**12**:1073-1081

[75] Bal S, Saha S. Scheming of microwave shielding effectiveness for X band considering functionalized MWNTs/epoxy composites. IOP Conference Series: Materials Science and Engineering. 2016;**115**(1):012027

[76] Arora M, Wahab MA, Saini P. Permittivity and electromagnetic interference shielding investigations of activated charcoal loaded acrylic coating compositions. Journal of Polymers. 2014

[77] Das NC, Chaki TK, Khastgir D, Chakraborty A. Electromagnetic interference shielding effectiveness of conductive carbon black and carbon fiber‐filled composites based on rubber and rubber blends. Advances in Polymer Technology: Journal of the Polymer Processing Institute. 2001;**20**(3):226-236

[78] Das NC, Chaki TK, Khastgir D, Chakraborty A. Electromagnetic interference shielding effectiveness of ethylene vinyl acetate based conductive composites containing carbon fillers. Journal of Applied Polymer Science. 2001;**80**(10):1601-1608

[79] Colaneri NF, Shacklette LW. IEEE Transactions on Instrumentation and Measurement. 1992;**41**:291

[80] Maxwell JC. New York: Oxford University Press; 2004. p. 828

[81] Van Uitert LG. The Journal of Chemical Physics. 1955;**23**:1883

Chapter 9

A New Boundary Element Formulation for Modeling and Optimization of Three-Temperature Nonlinear Generalized Magneto-Thermoelastic Problems of FGA Composite Microstructures

Mohamed Abdelsabour Fahmy

Abstract

The main purpose of this chapter is to propose a new boundary element formulation for the modeling and optimization of three-temperature nonlinear generalized magneto-thermoelastic functionally graded anisotropic (FGA) composite microstructures' problems, which is the gap of this study. Numerical results show that anisotropy and the functionally graded material have great influences on the nonlinear displacement sensitivities and nonlinear thermal stress sensitivities of composite microstructure optimization problem. Since, there are no available data for comparison, except for the problems with one-temperature heat conduction model, we considered the special case of our general study based on replacing three-temperature radiative heat conductions with one-temperature heat conduction. In the considered special case, numerical results demonstrate the validity and accuracy of the proposed technique. In order to solve the optimization problem, the method of moving asymptotes (MMA) based on the bi-evolutionary structural optimization method (BESO) has been implemented. A new class of composite microstructures problems with holes or inclusions was studied. The two-phase magneto-thermoelastic composite microstructure which is studied in this chapter consists of two different FGA materials. Through this chapter, we investigated that the optimal material distribution of the composite microstructures depends strongly on the heat conduction model, functionally graded parameter, and shapes of holes or inclusions.

Keywords: boundary element method, modeling and optimization, three-temperature, nonlinear generalized magneto-thermoelasticity, functionally graded anisotropic, composite microstructures

1. Introduction

In the last few years, there is significant interest in using advanced composite structures, and among the oldest examples of them, reinforced concrete, mixing concrete and steel, and plastics laminated with wood. The main benefit of the composite structures which consist of two or more different materials is that the properties of each material can be combined to form a single unit that performs better than the separate component parts. The most common form of a composite structure in construction is a steel and concrete composite, where concrete works well in pressure but has less resistance to tension. However, steel is extremely strong in tension, and when tied together, it results in a highly efficient and lightweight unit usually used for structures such as buildings and multistory bridges. Although fiberglass and carbon/epoxy composites are not yet as important as the oldest advanced composite structures in terms of tonnage or total revenue, they are very important in engineering, aerospace, transportation, bioengineering, optics, electronics, commodities, chemical plant, and energy industries, especially for the new airplanes that will concentrate on achieving major improvements in the fuel use, emissions, noise, transportation energy consumption, and other important issues to conserve the environment [1–21].

Microstructure has been known to play a major role in determining the behavior of material. Therefore, material engineers strive to control the microstructure by improving their properties with the aim of producing a uniform microstructure throughout the material. They also produced FGMs whose microstructures depend on the position by treating the microstructure as a position-dependent variable; the properties of different materials can be combined into one component to achieve an optimum performance in a specific application [22, 23].

In recent years, great attention has been directed toward the study of nonlinear generalized magneto-thermoelastic interactions in functionally graded anisotropic (FGA) structures due to its many applications in physics, geophysics, earthquake engineering, astronautics, aeronautics, mining engineering, military technologies, plasma, robotics, high-energy particle accelerators, nuclear reactors, automobile industries, nuclear plants, soil dynamics, and other engineering and industrial applications. Duhamel [24] and Neuman [25] proposed the classical thermoelasticity (CTE) theory which has the following two paradoxes: first, the infinite propagation speeds of thermal signals are predicted, and second, there is no any elastic term included in heat equation. Biot [26] invented the classical coupled thermoelasticity (CCTE) theory to beat the first paradox in CTE, but CTE and CCTE share the second paradox. Then, numerous generalized thermoelasticity theories have been introduced to overcome the two paradoxes inherent in CTE, such as the extended thermoelasticity (ETE) theory of Lord and Shulman [27]; temperature-rate-dependent thermoelasticity (TRDTE) theory of Green and Lindsay [28]; three linear generalized thermoelasticity theories of Green and Naghdi (GN) [29, 30]; namely I, II, and III, respectively [where, GN theory I is based on Fourier's law of heat conduction and is identical to CTE theory, GN theory II characterizes the thermoelasticity without energy dissipation (TEWOED), and GN theory III characterizes the thermoelasticity with energy dissipation (TEWED)]; dual phase-lag thermoelasticity (DPLTE) [31, 32]; and three-phase-lag thermoelasticity (TPLTE) [33].

A large amount of research has been done on the generalized problems of thermoelasticity [34–44]. Our interest in studying the three-temperature thermoelasticity [45–49] has increased due to its important low-temperature and high-temperature applications. Due to the computational difficulties, inherent in solving three-temperature nonlinear generalized magneto-thermoelastic problems of FGA composite microstructures, the problems become too complicated with no general analytical solution. Therefore, we propose a new boundary element modeling technique which has recently been successfully developed and implemented to obtain the approximate solutions for such problems. Now, the boundary element method (BEM), which is also called boundary integral equation method, has been widely adopted in a large variety of engineering and industrial applications. In the BEM, only the boundary of the solution domain needs to be discretized, so, it has a major advantage over other methods which require the whole domain discretization, such as the finite difference method (FDM) [50–52], discontinuous Galerkin method (DGM) [53], and finite element method (FEM) [54–57]. This advantage of BEM over domain methods has significant importance for modeling of nonlinear generalized thermoelastic problems which can be implemented using BEM with little cost and less input data [58–71]. Recently, scientists were convinced that only the FEM method could solve complex engineering problems. But now after the huge achievements of the BEM and its ability to solve complex engineering problems with high efficiency, it gets them to change their conviction. Also, they tried to combine FEM and BEM in the solution of their complex problems.

The main aim of this chapter is to propose a novel boundary element formulation for modeling and optimization of three-temperature nonlinear generalized thermoelastic problems of functionally graded anisotropic (FGA) composite microstructures. The proposed boundary element technique has been *implemented* successfully for solving several engineering, scientific and industrial applications due to its simplicity, efficiency, ease of use, and applicability [72–85]. The numerical results are presented graphically to show the influence of anisotropy and functionally graded materials on the sensitivities of displacements and thermal stresses. Also, numerical results show the effect of heat conduction model, functionally graded parameter, holes shape, and inclusions shape. Numerical results demonstrate the validity and accuracy of our proposed BEM formulation and technique.

A brief summary of the chapter is as follows: Section 1 introduces an overview of the historical background for a better understanding of the nonlinear generalized magneto-thermoelastic problems and composite materials applications. Section 2 describes the physical modeling of the three-temperature nonlinear generalized thermoelastic problems of FGA composite microstructures. Section 3 outlines the BEM implementation for solving the governing equations of the considered problem to obtain the three temperatures and displacement fields. Section 4 outlines the topology optimization technique used to obtain the optimal composite microstructure with and without holes or inclusions of various shapes. Section 5 presents the new numerical results that describe the effects of anisotropy and functionally graded parameters on the problem's fields' sensitivities during the optimization process. Section 6 outlines the significant findings of this chapter.

2. Formulation of the problem

Consider a Cartesian coordinates system $Ox_1x_2x_3$ as shown in **Figure 1**. We shall consider a functionally graded anisotropic composite microstructure of a finite thickness $\underline{\beta}$ placed in a primary magnetic field H_0 acting in the direction of the x_3-axis. The considered composite microstructure occupies the region $R = \left\{(x_1, x_2, x_3) : 0 < x_1 < \underline{\alpha}, 0 < x_2 < \underline{\beta}, 0 < x_3 < \underline{\gamma}\right\}$ with functionally graded material properties in the thickness direction.

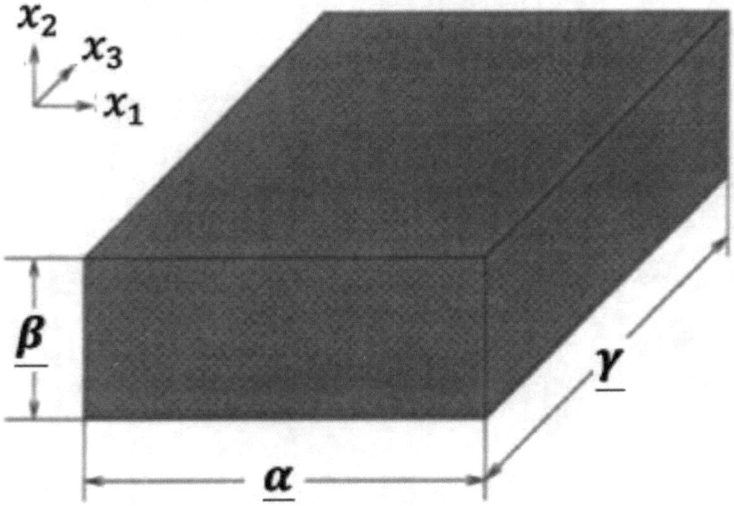

Figure 1.
Computational domain of considered structure.

The unified governing equations of three-temperature nonlinear generalized magneto-thermoelasticity for FGA composite microstructures can be expressed as follows [45–49]:

$$\sigma_{ab,b} + \tau_{ab,b} = \rho(x+1)^m \ddot{u}_a \tag{1}$$

$$\sigma_{ab} = (x+1)^m \left[C_{abfg} u_{f,g} - \beta_{ab}\left(T - T_0 + \tau_1 \dot{T}\right)\right] \tag{2}$$

$$\tau_{ab} = \mu(x+1)^m \left(\tilde{h}_a H_b + \tilde{h}_b H_a - \delta_{ba}\left(\tilde{h}_f H_f\right)\right) \tag{3}$$

The *2D-3T* radiative heat conduction Eqs. (7)–(9) can be expressed as follows:

$$\nabla\left[\left(\delta_{1j}\mathbb{K}_\alpha + \delta_{2j}\mathbb{K}_\alpha^*\right)\nabla T_\alpha(r,\tau)\right] - \overline{\mathbb{W}}(r,\tau) = C_{v\alpha}\rho(x+1)^m \delta_1 \delta_{1j} \frac{\partial T_\alpha(r,\tau)}{\partial \tau} \tag{4}$$

where

$$\overline{\mathbb{W}}(r,\tau) = \begin{cases} \rho\mathbb{W}_{ei}(T_e - T_i) + \rho\mathbb{W}_{er}\left(T_e - T_p\right) + \overline{\overline{\mathbb{W}}}, \alpha = e, \delta_1 = 1 & \\ -\rho\,\mathbb{W}_{ei}(T_e - T_i) + \overline{\overline{\mathbb{W}}}, & \alpha = i, \delta_1 = 1 \\ -\rho\,\mathbb{W}_{er}\left(T_e - T_p\right) + \overline{\overline{\mathbb{W}}}, & \alpha = p, \delta_1 = T_p^3 \end{cases}, C_{v\alpha} = \begin{cases} c_e & \alpha = e \\ c_i & \alpha = i \\ c_p T_p^3 & \alpha = p \end{cases} \tag{5}$$

in which

$$\begin{aligned} \overline{\overline{\mathbb{W}}}(r,\tau) = -\delta_{2j}\mathbb{K}_\alpha \dot{T}_{\alpha,ab} + \beta_{ab} T_{\alpha 0}(x+1)^m\left[\mathring{A}\delta_{1i}\dot{u}_{a,b} + (\tau_0 + \delta_{2i})\ddot{u}_{a,b}\right] \\ +\rho c_\alpha (x+1)^m\left[\left(\tau_0 + \delta_{1j}\tau_2 + \delta_{2j}\right)\ddot{T}_\alpha\right] \end{aligned} \tag{6}$$

and

$$\mathbb{W}_{ei} = \rho\mathbb{A}_{ei} T_e^{-2/3}, \mathbb{W}_{er} = \rho\mathbb{A}_{er} T_e^{-1/2}, \mathbb{K}_\alpha = \mathbb{A}_\alpha T_\alpha^{5/2}, \alpha = e, i, \mathbb{K}_p = \mathbb{A}_p T_p^{3+\mathbb{B}} \tag{7}$$

where σ_{ab}, τ_{ab}, u_k, T_α, and $T_{\alpha 0}$ are the mechanical stress tensor, Maxwell's electromagnetic stress tensor, displacement vector, temperature, and reference temperature, respectively; $C_{abfg}\left(C_{abfg} = C_{fgab} = C_{bafg}\right)$ and $\beta_{ab}\left(\beta_{ab} = \beta_{ba}\right)$ are, respectively, the constant elastic moduli and stress-temperature coefficients of the anisotropic medium; μ is the magnetic permeability; $\tilde{h}$ is the perturbed magnetic field; $\mathbb{K}_\alpha(\alpha = e, i, p)$ are the thermal conductivity coefficients; $C_{v\alpha}(\alpha = e, i, p)$ are specific heat coefficients; e, i, and p denote electron, ion, and phonon, respectively; $\mathbb{K}_\alpha^*$ is the second order tensor associated with the TEWED and TEWOED theories; $\mathbb{W}_{\text{ei}}$ is the electron-ion energy coefficient; $\mathbb{W}_{ep}$ is the electron-phonon energy coefficient; $c_\alpha(\alpha = e, i, p)$ are constants; ρ, τ, and Å are the density, time, and unified parameter which introduced to consolidate all theories into a unified equations system, respectively; τ_0, τ_1, and τ_2 are the relaxation times; and m is a functionally graded parameter. Also, g_1, g_2, Ψ_f, and δ_f are suitably prescribed functions; $\bar{t}_a$ are the tractions defined by $\bar{t}_a = \sigma_{ab} n_b$; and δ_{1j} and δ_{2j} are the Kronecker delta functions.

A superposed dot denotes the differentiation with respect to the time, and a comma followed by a subscript denotes partial differentiation with respect to the corresponding coordinates.

The unit mass total energy can be written as

$$P = P_e + P_i + P_p, P_e = c_e T_e, P_i = c_i T_i, P_p = \frac{1}{4} c_p T_p^4 \tag{8}$$

By using the following initial and boundary conditions:

$$T_\alpha(x, y, 0) = T_\alpha^0(x, y) = g_1(x, \tau) \tag{9}$$

$$\mathbb{K}_\alpha \frac{\partial T_\alpha}{\partial n}\bigg|_{C_1} = 0, \alpha = e, i, T_r\bigg|_{C_1} = g_2(x, \tau) \tag{10}$$

$$\mathbb{K}_\alpha \frac{\partial T_\alpha}{\partial n}\bigg|_{C_2} = 0, \alpha = e, i, p \tag{11}$$

$$u_f(x, y, 0) = \dot{u}_f(x, y, 0) = 0 \text{ for } (x, y) \in R \cup C \tag{12}$$

$$u_f(x, y, \tau) = \Psi_f(x, y, \tau) \text{ for } (x, y) \in C_3 \tag{13}$$

$$\bar{t}_a(x, y, \tau) = \delta_f(x, y, \tau) \text{ for } (x, y) \in C_4, \tau > 0, C = C_3 \cup C_4, C_3 \cap C_4 = \varnothing \tag{14}$$

By using the fundamental solution that satisfies the following equation:

$$D\nabla^2 T_\alpha + \frac{\partial T_\alpha^*}{\partial n} = -\delta(r - p_i)\delta(\tau - r), D = \frac{\mathbb{K}_\alpha}{\rho c} \tag{15}$$

where p_i are singular points.

The above governing Eqs. (1)–(4) can be reduced to the different theories of three-temperature nonlinear generalized magneto-thermoelasticity for FGA composite microstructures as follows [77]:

$$\text{CTE}: j = 1, \text{Å} = 0 \text{ and } \tau_0 = \tau_1 = \tau_2 = 0 \tag{16}$$

$$\text{CCTE}: j = 1, \text{Å} = 1 \text{ and } \tau_0 = \tau_1 = \tau_2 = 0 \tag{17}$$

$$\text{ETE}: j = 1, \text{Å} = 1 \text{ and } \tau_1 = \tau_2 = 0 \tag{18}$$

$$\text{TRDTE}: j = 1, \text{Å} = 1 \text{ and } \tau_0 = 0 \tag{19}$$

$$\text{TEWED}: j = 2, \text{Å} = 0 \text{ and } \tau_0 = 0 \tag{20}$$

$$\text{TEWOED}: j = 2, \text{Å} = 0, \tau_0 = 0 \text{ and } \mathbb{K}_\alpha \to 0 \tag{21}$$

3. BEM implementation

By using Eqs. (2) and (3), we can write Eq. (1) as follows:

$$L_{gb}u_f = \rho\ddot{u}_a - D_aT = f_{gb} \tag{22}$$

where

$$L_{gb} = D_{abf}\frac{\partial}{\partial x_b} + D_{af} + \Lambda D_{a1f}, D_{abf} = C_{abfg}\varepsilon, \varepsilon = \frac{\partial}{\partial x_g},$$

$$D_{af} = \mu H_0^2\left(\frac{\partial}{\partial x_a} + \delta_{a1}\Lambda\right)\frac{\partial}{\partial x_f}, D_a = -\beta_{ab}\left(\frac{\partial}{\partial x_b} + \delta_{b1}\Lambda + \tau_1\left(\frac{\partial}{\partial x_b} + \Lambda\right)\frac{\partial}{\partial \tau}\right),$$

$$\Lambda = \frac{m}{x+1}, f_{gb} = \rho\ddot{u}_a - D_aT. \tag{23}$$

The field equations can be written in the following operator form:

$$L_{gb}u_f = f_{gb} \tag{24}$$

$$L_{ab}T = f_{ab} \tag{25}$$

where the operators L_{gb} and f_{gb} are defined above in Eq. (23), and the operators L_{ab} and f_{ab} are defined as follows:

$$L_{ab} = \nabla\left(\delta_{2j}\mathbb{K}_\alpha^*\right)\nabla \tag{26}$$

$$f_{ab} = -\nabla\left(\delta_{1j}\mathbb{K}_\alpha\right)\nabla + c_\alpha\rho\delta_1\delta_{1j}(x+1)^m\frac{\partial T_\alpha(r,\tau)}{\partial \tau} + \overline{\mathbb{W}}(r,\tau) \tag{27}$$

By applying the weighted residual method (WRM) to the differential Eq. (24), we obtain

$$\int_R\left(L_{gb}u_f - f_{gb}\right)u_{da}^*dR = 0 \tag{28}$$

Now, we can choose the fundamental solution u_{df}^* as weighting function as

$$L_{gb}u_{df}^* = -\delta_{ad}\delta(x,\xi) \tag{29}$$

The corresponding traction field can be expressed as

$$t_{da}^* = C_{abfg}(x+1)^m u_{df,g}^* n_b \tag{30}$$

The traction vector can be expressed as

$$t_a = \frac{\bar{t}_a}{(x+1)^m} = (x+1)^m\left(C_{abfg}u_{f,g} - \beta_{ab}\left(T + \tau_1\dot{T}\right)\right)n_b \tag{31}$$

By applying integration by parts to Eq. (28) and using the sifting property with Eqs. (29) and (31), we obtain

$$u_d(\xi) = \int_C \left(u^*_{da} t_a - t^*_{da} u_a + u^*_{da} \beta_{ab} (x+1)^m T n_b\right) dC - \int_R f_{gb} u^*_{da} dR \tag{32}$$

The fundamental solution T^* can be expressed as

$$L_{ab} T^* = -\delta(x, \xi) \tag{33}$$

By implementing the WRM and integration by parts, we can write Eq. (25) in the following form:

$$\int_R (L_{ab} T T^* - L_{ab} T^* T) dR = \int_C (q^* T - q T^*) dC \tag{34}$$

where

$$q = -\mathbb{K}_a t_{,b} n_a \tag{35}$$

$$q^* = -\mathbb{K}_a T^*_{,b} n_a \tag{36}$$

Based on the sifting property, we can express Eq. (34) as follows:

$$T(\xi) = \int_C (q^* T - q T^*) dC - \int_R f_{ab} T^* dR \tag{37}$$

The field Eqs. (32) and (37) can be written in one equation of the form:

$$\begin{bmatrix} u_d(\xi) \\ T(\xi) \end{bmatrix} = \int_C \left\{ -\begin{bmatrix} t^*_{da} & -u^*_{da} \beta_{ab} (x+1)^m n_b \\ 0 & -q^* \end{bmatrix} \begin{bmatrix} u_a \\ T \end{bmatrix} + \begin{bmatrix} u^*_{da} & 0 \\ 0 & -T^* \end{bmatrix} \begin{bmatrix} t_a \\ q \end{bmatrix} \right\} dC - \int_R \begin{bmatrix} u^*_{da} & 0 \\ 0 & -T^* \end{bmatrix} \begin{bmatrix} f_{gb} \\ -f_{ab} \end{bmatrix} dR \tag{38}$$

The generalized thermoelastic vectors and tensors can be written in contracted notation as follows:

$$U_A = \begin{cases} u_a & a = A = 1, 2, 3 \\ T & A = 4 \end{cases} \tag{39}$$

$$T_A = \begin{cases} t_a & a = A = 1, 2, 3 \\ q & A = 4 \end{cases} \tag{40}$$

$$U^*_{DA} = \begin{cases} u^*_{da} & d = D = 1, 2, 3; a = A = 1, 2, 3 \\ 0 & d = D = 1, 2, 3; A = 4 \\ 0 & D = 4; a = A = 1, 2, 3 \\ -T^* & D = 4; A = 4 \end{cases} \tag{41}$$

$$\tilde{T}^*_{DA} = \begin{cases} t^*_{da} & d = D = 1,2,3; a = A = 1,2,3 \\ -u^*_d & d = D = 1,2,3; A = 4 \\ 0 & D = 4; a = A = 1,2,3 \\ -q^* & D = 4; A = 4 \end{cases} \tag{42}$$

$$\tilde{u}^*_d = u^*_{da}\beta_{af}n_f \tag{43}$$

The thermoelastic representation formula (38) can be written in contracted notation as

$$U_D(\xi) = \int_C (U^*_{DA}\mathrm{T}_A - \tilde{T}_{DA}U_A)dC - \int_R U^*_{DA}S_A dR \tag{44}$$

The vector S_A can be splitted as

$$S_A = S^T_A + S^{\dot{T}}_A + S^{\ddot{T}}_A + S^{\dot{u}}_A + S^{\ddot{u}}_A \tag{45}$$

where $S^T_A = \omega_{AF}U_F$
with

$$\omega_{AF} = \begin{cases} -D_a & A = 1,2,3; F = 4 \\ \nabla(\delta_{2j}\mathbb{K}^*_\alpha)\nabla T_\alpha + \begin{cases} \rho\mathbb{W}_{ei}(T_e - T_i) + \rho\mathbb{W}_{er}(T_e - T_p), & \alpha = e, \delta_1 = 1 \\ -\rho\mathbb{W}_{ei}(T_e - T_i), & \alpha = i, \delta_1 = 1 \\ -\rho\mathbb{W}_{er}(T_e - T_p), & \alpha = p, \delta_1 = \frac{4}{\rho}T^3_p \end{cases} & \text{otherwise} \end{cases} \tag{46}$$

$$S^{\dot{T}}_A = \left(-\delta_{2j}\mathbb{K}_\alpha \frac{\partial}{\partial x_a}\frac{\partial}{\partial x_b} + c_\alpha\rho\delta_1\delta_{1j}(x+1)^m\right)\delta_{AF}\dot{U}_F$$

$$\text{with } \delta_{AF} = \begin{cases} 1 & A = 4; F = 4 \\ 0 & \text{otherwise} \end{cases} \tag{47}$$

$$S^{\ddot{T}}_A = -\rho c_\alpha (x+1)^m(\tau_0 + \delta_{1j}\tau_2 + \delta_{2j})\delta_{AF}\ddot{U}_F \tag{48}$$

$$S^{\dot{u}}_A = -\beta_{ab}(x+1)^m T_{\alpha 0}\text{Å}\delta_{1j}\dot{U}_F \tag{49}$$

$$S^{\ddot{u}}_A = Ⅎ\ddot{U}_F \text{ with } Ⅎ = \begin{cases} \rho(x+1)^m & A = 1,2,3; \mathrm{F} = 1,2,3, \\ -T_{\alpha 0}\beta_{fg}(x+1)^m(\tau_0 + \delta_{2j})A = 4; f = F = 4 \end{cases} \tag{50}$$

The thermoelastic representation formula (38) can also be expressed as follows:

$$[S_A] = \begin{bmatrix} -D_a T_\alpha \\ \nabla(\delta_{2j}\mathbb{K}^*_\alpha)\nabla T_\alpha + \begin{cases} \rho\mathbb{W}_{ei}(T_e - T_i) + \rho\mathbb{W}_{er}(T_e - T_p), & \alpha = e, \delta_1 = 1 \\ -\rho\mathbb{W}_{ei}(T_e - T_i), & \alpha = i, \delta_1 = 1 \\ -\rho\mathbb{W}_{er}(T_e - T_p), & \alpha = p, \delta_1 = \frac{4}{\rho}T^3_p \end{cases} \end{bmatrix}$$

$$+\left(\delta_{2j}\mathbb{K}_\alpha\frac{\partial}{\partial x_a}\frac{\partial}{\partial x_b}-c_\alpha\rho\delta_1\delta_{1j}(x+1)^m\right)\begin{bmatrix}0\\ \dot{T}_\alpha\end{bmatrix}$$

$$-\rho c_\alpha(x+1)^m(\tau_0+\delta_{1j}\tau_2+\delta_{2j})\begin{bmatrix}0\\ \ddot{T}_\alpha\end{bmatrix}-\beta_{ab}(x+1)^m T_{\alpha 0}\mathring{A}\delta_{1j}\begin{bmatrix}0\\ \dot{u}_{fg}\end{bmatrix} \tag{51}$$

$$+\begin{bmatrix}\rho(x+1)^m\ddot{u}_a\\ -T_{\alpha 0}\beta_{fg}(x+1)^m(\tau_0+\delta_{2j})\ddot{u}_{fg}\end{bmatrix}$$

In order to transform the domain integral in Eq. (44) to the boundary, we approximate the source vector S_A by a series of given known functions f^q_{AE} and unknown coefficients α^q_E:

$$S_A\approx\sum_{q=1}^{N}f^q_{AN}\alpha^q_N \tag{52}$$

Thus, the thermoelastic representation formula (44) can be expressed as

$$U_D(\xi)=\int_C\left(U^*_{DA}T_A-\tilde{T}^*_{DA}U_A\right)dC-\sum_{q=1}^{N}\int_R U^*_{DA}f^q_{AN}dR\alpha^q_N \tag{53}$$

By applying the WRM to the following elastic and thermal equations:

$$L_{gb}u^q_{fn}=f^q_{an} \tag{54}$$

$$L_{ab}T^q=f^q_{pj} \tag{55}$$

Now, the weighting functions were chosen as the elastic and thermal fundamental solutions u^*_{da} and T^*.

Then, the representation formulae of elastic and thermal fields are given as follows:

$$u^q_{dn}(\xi)=\int_C\left(u^*_{da}t^q_{an}-t^*_{da}u^q_{an}\right)dC-\int_R u^*_{da}f^q_{an}dR \tag{56}$$

$$T^q(\xi)=\int_C\left(q^*T^q-q^qT^*\right)dC-\int_R f^qT^*dR \tag{57}$$

The elastic and thermal representation formulae can be combined in one single equation as

$$U^q_{DN}(\xi)=\int_C\left(U^*_{DA}T^q_{AN}-T^*_{DA}U^q_{AN}\right)dC-\int_R U^*_{DA}f^q_{AN}dR \tag{58}$$

By substituting from Eq. (58) into Eq. (53), the coupled thermoelastic representation formula can be expressed as follows:

$$U_D(\xi) = \int_C \left(U^*_{DA} T_A - \breve{T}^*_{DA} U_A \right) dC + \sum_{q=1}^{N} \left(U^q_{DN}(\xi) + \int_C \left(T^*_{DA} U^q_{AN} - U^*_{DA} T^q_{AN} \right) dC \right) \alpha^q_N \tag{59}$$

By differentiation of Eq. (59) with respect to ξ_l, we obtain

$$\frac{\partial U_D(\xi)}{\partial \xi_l} = -\int_C \left(U^*_{DA,l} T_A - \breve{T}^*_{DA,l} U_A \right) dC + \sum_{q=1}^{N} \left(\frac{\partial U^q_{DN}(\xi)}{\partial \xi_l} - \int_C \left(T^*_{DA,l} U^q_{AN} - U^*_{DA,l} T^q_{AN} \right) dC \right) \alpha^q_N \tag{60}$$

According to the procedure described in Fahmy [78], the boundary integral Eq. (59) can be expressed as

$$\breve{\zeta} U - \eta T = \left(\zeta \breve{U} - \eta \wp \right) \alpha \tag{61}$$

According to the technique of Partridge et al. [68], the displacements U_F and velocities $\dot{U}_F$ can be approximated as

$$U_F \approx \sum_{q=1}^{N} f^q_{FD}(x) \gamma^q_D \tag{62}$$

$$\dot{U}_F \approx \sum_{q=1}^{N} f^q_{FD}(x) \tilde{\gamma}^q_D \tag{63}$$

where f^q_{FD} are known functions, and γ^q_D and $\tilde{\gamma}^q_D$ are unknown coefficients. The gradients of the displacement and velocity can be approximated as

$$U_{F,g} \approx \sum_{q=1}^{N} f^q_{FD,g}(x) \gamma^q_K \tag{64}$$

$$\dot{U}_{F,g} \approx \sum_{q=1}^{N} f^q_{FD,g}(x) \tilde{\gamma}^q_D \tag{65}$$

By substituting from Eqs. (62) and (63) into Eqs. (46) and (49), the corresponding source terms can be expressed as

$$S^T_A = \sum_{q=1}^{N} S^{T,q}_{AD} \gamma^q_D \tag{66}$$

$$S^{\dot{u}}_A = -\beta_{ab} (x+1)^m T_{a0} \mathring{A} \delta_{1j} \sum_{q=1}^{N} S^{\dot{u},q}_{AD} \tilde{\gamma}^q_D \tag{67}$$

where

$$S_{AD}^{T,q} = S_{AF} f_{FD,q}^{q} \tag{68}$$

$$S_{AD}^{\dot{u},q} = S_{FA} f_{FD,g}^{q} \tag{69}$$

By applying the point collocation procedure of Gaul et al. [10] to Eqs. (52), (62), and (63), we obtain the following equation system:

$$\check{S} = J\alpha, U = J'\gamma, \dot{U} = J'\tilde{\gamma} \tag{70}$$

Solving the system (70) for α, γ, and $\tilde{\gamma}$ yields

$$\alpha = J^{-1}\check{S}\gamma = J'^{-1}U \ \tilde{\gamma} = J'^{-1}\dot{U} \tag{71}$$

Now, we can write the coefficients α in terms of nodal values of the displacements, U, velocities, $\dot{U}$, and accelerations, $\ddot{U}$ as follows:

$$\begin{aligned}\alpha = J^{-1}\big(&\check{S}^0 + B^T J'^{-1}U \\ &+\left[\left(\delta_{2j}\mathbb{K}_\alpha \frac{\partial}{\partial x_a}\frac{\partial}{\partial x_b} - c_\alpha\rho\delta_1\delta_{1j}(x+1)^m\right)\delta_{AF} - \beta_{ab}(x\right. \\ &\left.+1)^m T_{\alpha 0}\mathring{A}\delta_{1j}J'^{-1}\right]\dot{U} \\ &+\left[-c_\alpha\rho(x+1)^m\left(\tau_0 + \delta_{1j}\tau_2 + \delta_{2j}\right)\delta_{AF}\right]\ddot{U}\big)\end{aligned} \tag{72}$$

By substituting from Eq. (72) into Eq. (61) and implementing implicit-implicit staggered algorithm of Farhat et al. [86], the governing equations can be rewritten as

$$\widehat{M}\ddot{U} + \widehat{\Gamma}\dot{U} + \widehat{K}U = \widehat{\mathbb{Q}} \tag{73}$$

$$\widehat{\mathrm{X}}\ddot{T} + \widehat{\mathrm{A}}\dot{T} + \widehat{\mathrm{B}}T = \widehat{\mathbb{Z}}\ddot{U} + \widehat{\mathbb{R}}\dot{U} \tag{74}$$

where

$$\begin{aligned}&V = (\eta\check{\wp} - \zeta\check{U})J^{-1}, \widehat{M} = V\tilde{A}, \widehat{\mathrm{X}} = -\rho c_\alpha(x+1)^m\left(\tau_0 + \delta_{1j}\tau_2 + \delta_{2j}\right), \\ &\widehat{K} = \tilde{\zeta} + VB^T J'^{-1}, \mathbb{Q} = \eta T + V\breve{S}^0, \widehat{\mathrm{B}} = \delta_{1j}\mathbb{K}_\alpha + \delta_{2j}\mathbb{K}_\alpha^*, \\ &\widehat{\Gamma} = V\left[\left(\mathbb{K}_\alpha\frac{\partial}{\partial x_a}\frac{\partial}{\partial x_b} - c_\alpha\rho(x+1)^m\delta_{1j}\right)\delta_{AF} - T_0\mathring{A}\delta_{1j}\beta_{fg}(x+1)^m J'^{-1}\right], \\ &\widehat{\mathbb{R}} = T_0\beta_{ab}(x+1)^m\mathring{A}\delta_{1j}, \widehat{\mathbb{Z}} = T_{\alpha 0}\beta_{ab}(x+1)^m\left(\tau_0 + \delta_{2j}\right), \\ &\widehat{\mathrm{A}} = \left(\delta_{2j}\mathbb{K}_\alpha^*\frac{\partial}{\partial x_a}\frac{\partial}{\partial x_b} - \rho c_\alpha(x+1)^m\delta_{1j}\right)\delta_{AF}.\end{aligned} \tag{75}$$

where V, $\widehat{M}$, $\widehat{\Gamma}$, $\widehat{K}$, $\widehat{\mathrm{A}}$, and $\widehat{\mathrm{B}}$ are represent the volume, mass, damping, stiffness, capacity, and conductivity matrices, respectively; $\ddot{U}$, $\dot{U}$, U, T, and $\widehat{\mathbb{Q}}$ represent the acceleration, velocity, displacement, temperature, and external force vectors, respectively, $\widehat{X}$ is a Green and Lindsay material constants vector, and $\widehat{\mathbb{Z}}$ and $\widehat{\mathbb{R}}$ are coupling matrices.

Hence, the governing equations lead to the following coupled system of differential-algebraic equations (DAEs) as in Farhat et al. [86]:

$$\widehat{M}\ddot{U}_{n+1}+\widehat{\Gamma}\dot{U}_{n+1}+\widehat{K}U_{n+1}=\widehat{\mathbb{Q}}^{p}_{n+1} \tag{76}$$

$$\widehat{X}\ddot{T}_{n+1}+\widehat{A}\dot{T}_{n+1}+\widehat{B}T_{n+1}=\widehat{\mathbb{Z}}\ddot{U}_{n+1}\widehat{\mathbb{R}}\dot{U}_{n+1} \tag{77}$$

where $\widehat{\mathbb{Q}}^{p}_{n+1}=\eta T^{p}_{n+1}+V\check{S}^{0}$ and T^{p}_{n+1}.

By integrating Eq. (73) and using Eq. (76), we get

$$\begin{aligned}\dot{U}_{n+1}&=\dot{U}_{n}+\frac{\Delta\tau}{2}\left(\ddot{U}_{n+1}+\ddot{U}_{n}\right)\\&=\dot{U}_{n}+\frac{\Delta\tau}{2}\left[\ddot{U}_{n}+\widehat{M}^{-1}\left(\widehat{\mathbb{Q}}^{p}_{n+1}-\widehat{\Gamma}\dot{U}_{n+1}-\widehat{K}U_{n+1}\right)\right]\end{aligned} \tag{78}$$

$$\begin{aligned}U_{n+1}&=U_{n}+\frac{\Delta\tau}{2}\left(\dot{U}_{n+1}+\dot{U}_{n}\right)\\&=U_{n}+\Delta\tau\dot{U}_{n}+\frac{\Delta\tau^{2}}{4}\left[\ddot{U}_{n}+\widehat{M}^{-1}\left(\widehat{\mathbb{Q}}^{p}_{n+1}-\widehat{\Gamma}\dot{U}_{n+1}-\widehat{K}U_{n+1}\right)\right]\end{aligned} \tag{79}$$

From Eq. (78) we obtain

$$\dot{U}_{n+1}=\overline{\gamma}^{-1}\left[\dot{U}_{n}+\frac{\Delta\tau}{2}\left[\ddot{U}_{n}+\widehat{M}^{-1}\left(\widehat{\mathbb{Q}}^{p}_{n+1}-\widehat{K}U_{n+1}\right)\right]\right] \tag{80}$$

where $\overline{\gamma}=\left(I\frac{\Delta\tau}{2}\widehat{M}^{-1}\widehat{\Gamma}\right)$.

Substitution of Eq. (80) in Eq. (79), we obtain

$$\begin{aligned}U_{n+1}=&\;U_{n}+\Delta\tau\dot{U}_{n}\\&+\frac{\Delta\tau^{2}}{4}\left[\ddot{U}_{n}+\widehat{M}^{-1}\left(\widehat{\mathbb{Q}}^{p}_{n+1}-\widehat{\Gamma}\,\overline{\gamma}^{-1}\left[\dot{U}_{n}+\frac{\Delta\tau}{2}\left[\ddot{U}_{n}+\widehat{M}^{-1}\left(\widehat{\mathbb{Q}}^{p}_{n+1}-\widehat{K}U_{n+1}\right)\right]\right]-\widehat{K}U_{n+1}\right)\right]\end{aligned} \tag{81}$$

Substituting $\dot{U}_{n+1}$ from Eq. (80) into Eq. (76), we obtain

$$\ddot{U}_{n+1}=\widehat{M}^{-1}\left[\widehat{\mathbb{Q}}^{p}_{n+1}-\widehat{\Gamma}\left[\overline{\gamma}^{-1}\left[\dot{U}_{n}+\frac{\Delta\tau}{2}\left[\ddot{U}_{n}+\widehat{M}^{-1}\left(\widehat{\mathbb{Q}}^{p}_{n+1}-\widehat{K}U_{n+1}\right)\right]\right]\right]-\widehat{K}U_{n+1}\right] \tag{82}$$

Integrating the heat Eq. (74) using the trapezoidal rule and Eq. (77), we get

$$\begin{aligned}\dot{T}_{n+1}&=\dot{T}_{n}+\frac{\Delta\tau}{2}\left(\ddot{T}_{n+1}+\ddot{T}_{n}\right)\\&=\dot{T}_{n}+\frac{\Delta\tau}{2}\left(\widehat{X}^{-1}\left[\widehat{\mathbb{Z}}\ddot{U}_{n+1}+\widehat{\mathbb{R}}\dot{U}_{n+1}-\widehat{A}\dot{T}_{n+1}-\widehat{B}T_{n+1}\right]+\ddot{T}_{n}\right)\end{aligned} \tag{83}$$

$$T_{n+1} = T_n + \frac{\Delta\tau}{2}\left(\dot{T}_{n+1} + \dot{T}_n\right)$$
$$= T_n + \Delta\tau\dot{T}_n + \frac{\Delta\tau^2}{4}\left(\ddot{T}_n + \widehat{\mathrm{X}}^{-1}\left[\widehat{\mathbb{Z}}\,\ddot{U}_{n+1} + \widehat{\mathbb{R}}\,\dot{U}_{n+1} - \widehat{\mathrm{A}}\,\dot{T}_{n+1} - \widehat{\mathrm{B}}\,T_{n+1}\right]\right) \tag{84}$$

From Eq. (83), we have

$$\dot{T}_{n+1} = \gamma^{-1}\left[\dot{T}_n + \frac{\Delta\tau}{2}\left(\widehat{\mathrm{X}}^{-1}\left[\widehat{\mathbb{Z}}\,\ddot{U}_{n+1} + \widehat{\mathbb{R}}\,\dot{U}_{n+1} - \widehat{\mathrm{B}}\,T_{n+1}\right] + \ddot{T}_n\right)\right] \tag{85}$$

where $\gamma = \left(I + \frac{1}{2}\widehat{\mathrm{A}}\,\Delta\tau\,\widehat{\mathrm{X}}^{-1}\right)$.

On substitution of Eq. (85) in Eq. (84), we obtain

$$T_{n+1} = T_n + \Delta\tau\dot{T}_n + \frac{\Delta\tau^2}{4}\left(\ddot{T}_n + \widehat{\mathrm{X}}^{-1}\left[\widehat{\mathbb{Z}}\,\ddot{U}_{n+1} + \widehat{\mathbb{R}}\,\dot{U}_{n+1}\right.\right.$$
$$\left.\left. - \widehat{\mathrm{A}}\left(\gamma^{-1}\left[\dot{T}_n + \frac{\Delta\tau}{2}\left(\widehat{\mathrm{X}}^{-1}\left[\widehat{\mathbb{Z}}\,\ddot{U}_{n+1} + \widehat{\mathbb{R}}\,\dot{U}_{n+1} - \widehat{\mathrm{B}}\,T_{n+1}\right] + \ddot{T}_n\right)\right]\right) - \widehat{\mathrm{B}}\,T_{n+1}\right]\right) \tag{86}$$

On substitution of $\dot{T}_{n+1}$ from Eq. (85) in Eq. (77), we get

$$\ddot{T}_{n+1} = \widehat{\mathrm{X}}^{-1}\left[\widehat{\mathbb{Z}}\,\ddot{U}_{n+1} + \widehat{\mathbb{R}}\,\dot{U}_{n+1}\right.$$
$$\left. - \widehat{\mathrm{A}}\left(\gamma^{-1}\left[\dot{T}_n + \frac{\Delta\tau}{2}\left(\widehat{\mathrm{X}}^{-1}\left[\widehat{\mathbb{Z}}\,\ddot{U}_{n+1} + \widehat{\mathbb{R}}\,\dot{U}_{n+1} - \widehat{\mathrm{B}}\,T_{n+1}\right] + \ddot{T}_n\right)\right]\right) - \widehat{\mathrm{B}}\,T_{n+1}\right]\Big) \tag{87}$$

Now, our algorithm for the solution of Eqs. (81) and (86) is obtained as follows:

First step. Predict the displacement field: $U^p_{n+1} = U_n$.

Second step. Substituting for $\dot{U}_{n+1}$ from Eq. (78) and substituting for $\ddot{U}_{n+1}$ from Eq. (76). Then, by using the resulted equations in Eq. (86) to obtain the temperature field.

Third step. Correct the displacement field (81) by using the computed temperature.

Fourth step. Compute $\dot{U}_{n+1}$, $\ddot{U}_{n+1}$, $\dot{T}_{n+1}$, and $\ddot{T}_{n+1}$ from Eqs. (80), (82), (85), and (87), respectively.

4. Design sensitivity and optimization

According to Fahmy [77, 78], the design sensitivities of the nonlinear temperature field and nonlinear displacement field can be performed by the implicit differentiation of Eqs. (76) and (77), respectively, which describe the structural response with respect to the design variables, then we can compute the nonlinear thermal stresses sensitivities.

In order to solve our topology optimization problem, the method of moving asymptotes (MMA) [87] has been implemented as an optimizer in our topology optimization program. The benefit of MMA algorithm is that it replaces the original nonlinear, non-convex optimization problem by a sequence of approximating convex subproblems which are much easier to solve. The implemented MMA is based on the bi-directional evolutionary structural optimization (BESO), which is the evolutionary topology optimization approach that allows modification of the structure by either adding efficient material or removing inefficient material to or from the structure design [88–96]. This addition or removal depends upon the sensitivity analysis. Sensitivity analysis is the estimation of the response of the structure to the modification of the input design variables and is dependent upon the calculation of derivatives.

The homogenized vector of thermal expansion coefficients α^H can be written in terms of the homogenized elastic matrix D^H and homogenized stress-temperature coefficients vector β^H as follows:

$$\alpha^H = \left(D^H\right)^{-1}\beta^H \tag{88}$$

For the material design, the derivative of the homogenized thermal expansion coefficients vector can be expressed as

$$\frac{\partial \alpha^H}{\partial X^m_{kl}} = \left(D^H\right)^{-1}\left(\frac{\partial \beta^H}{\partial X^m_{kl}} - \frac{\partial D^H}{\partial X^m_{kl}}\alpha^H\right) \tag{89}$$

where $\frac{\partial D^H}{\partial X^m_{kl}}$ and $\frac{\partial \beta^H}{\partial X^m_{kl}}$ for any lth material phase, can be calculated using the adjoint variable method [91] as

$$\frac{\partial D^H}{\partial X^m_{kl}} = \frac{1}{|\Omega|}\int_Y (I - B^m U^m)^T \frac{\partial D^m}{\partial X^m_{kl}} (I - B^m U^m)\, dy \tag{90}$$

and

$$\begin{aligned}\frac{\partial \beta^H}{\partial X^m_{kl}} &= \frac{1}{|\Omega|}\int_Y (I - B^m U^m)^T \frac{\partial D^m}{\partial X^m_{kl}} (\alpha^m - B^m \varphi^m)\, dy \\ &+ \frac{1}{|\Omega|}\int_Y (I - B^m U^m)^T D^m \frac{\partial \alpha^m}{\partial X^m_{kl}} dy\end{aligned} \tag{91}$$

where, $|\Omega|$ is the volume of the base cell.

5. Numerical examples, results, and discussion

The proposed technique used in the current chapter should be applicable to any three-temperature nonlinear generalized magneto-thermoelastic problem. The application is for the purpose of illustration.

The two anisotropic materials considered in the calculation are monoclinic graphite-epoxy and North Sea sandstone reservoir rock, where the physical data of monoclinic graphite-epoxy material is given as follows:

Elasticity tensor:

$$C_{pjkl} = \begin{bmatrix} 430.1 & 130.4 & 18.2 & 0 & 0 & 201.3 \\ 130.4 & 116.7 & 21.0 & 0 & 0 & 70.1 \\ 18.2 & 21.0 & 73.6 & 0 & 0 & 2.4 \\ 0 & 0 & 0 & 19.8 & -8.0 & 0 \\ 0 & 0 & 0 & -8.0 & 29.1 & 0 \\ 201.3 & 70.1 & 2.4 & 0 & 0 & 147.3 \end{bmatrix} \text{GPa} \tag{92}$$

Mechanical temperature coefficient:

$$\beta_{pj} = \begin{bmatrix} 1.01 & 2.00 & 0 \\ 2.00 & 1.48 & 0 \\ 0 & 0 & 7.52 \end{bmatrix} \cdot 10^6 \mathrm{N/Km^2} \tag{93}$$

Tensor of thermal conductivity:

$$k_{pj} = \begin{bmatrix} 5.2 & 0 & 0 \\ 0 & 7.6 & 0 \\ 0 & 0 & 38.3 \end{bmatrix} \mathrm{W/km} \tag{94}$$

Mass density $\rho = 7820\,\mathrm{kg/m^3}$ and heat capacity $c = 461\ \mathrm{J/(kg{\cdot}K)}$, $H_0 = 1000000$ Oersted, $\mu = 0.5$ Gauss/Oersted, $h = 2$, and $\Delta\tau = 0.0001$.

The physical data of the North Sea sandstone reservoir rock is given as follows:

Elasticity tensor:

$$C_{pjkl} = \begin{bmatrix} 17.77 & 3.78 & 3.76 & 0.24 & -0.28 & 0.03 \\ 3.78 & 19.45 & 4.13 & 0 & 0 & 1.13 \\ 3.76 & 4.13 & 21.79 & 0 & 0 & 0.38 \\ 0 & 0 & 0 & 8.30 & 0.66 & 0 \\ 0 & 0 & 0 & 0.66 & 7.62 & 0 \\ 0.03 & 1.13 & 0.38 & 0 & 0 & 7.77 \end{bmatrix} \mathrm{GPa} \tag{95}$$

Mechanical temperature coefficient:

$$\beta_{pj} = \begin{bmatrix} 0.001 & 0.02 & 0 \\ 0.02 & 0.006 & 0 \\ 0 & 0 & 0.05 \end{bmatrix} \cdot 10^6\ \mathrm{N/Km^2} \tag{96}$$

Tensor of thermal conductivity:

$$k_{pj} = \begin{bmatrix} 1 & 0.1 & 0.2 \\ 0.1 & 1.1 & 0.15 \\ 0.2 & 0.15 & 0.9 \end{bmatrix} \mathrm{W/km} \tag{97}$$

Mass density $\rho = 2216\mathrm{kg/m^3}$ and heat capacity $c = 0.1\ \mathrm{J/(kg{\cdot}K)}$, $H_0 = 1000000$ Oersted, $\mu = 0.5$ Gauss/Oersted, $h = 2$, and $\Delta\tau = 0.0001$.

The initial and boundary conditions considered in the calculations are

$$\text{at } \tau = 0 \ u_1 = u_2 = \dot{u}_1 = \dot{u}_2 = 0, T = 0 \tag{98}$$

$$\text{at } x = 0 \ \frac{\partial u_1}{\partial x} = \frac{\partial u_2}{\partial x} = 0, \frac{\partial T}{\partial x} = 0 \tag{99}$$

$$\text{at } x = h \ \frac{\partial u_1}{\partial x} = \frac{\partial u_2}{\partial x} = 0, \frac{\partial T}{\partial x} = 0 \tag{100}$$

$$\text{at } y = 0 \ \frac{\partial u_1}{\partial y} = \frac{\partial u_2}{\partial y} = 0, \frac{\partial T}{\partial y} = 0 \tag{101}$$

$$\text{at } y = b\,\frac{\partial u_1}{\partial y} = \frac{\partial u_2}{\partial y} = 0,\ \frac{\partial T}{\partial y} = 0 \tag{102}$$

In order to study the effects of anisotropy and functionally graded materials on composite microstructure, we consider the following four cases, namely, isotropic homogeneous (IH), isotropic functionally graded (IF), anisotropic homogeneous (AH), and anisotropic functionally graded (AF). Also, we considered total temperature $T(T = T_e + T_i + T_p)$ as the considered temperature field in all calculations of this study.

Figure 2 shows the variations of the nonlinear three-temperature T_e, T_i, and T_p and total temperature T $(T = T_e + T_i + T_p)$, with the time τ through composite microstructure.

Figures 3 and **4** show the variation of the nonlinear displacement sensitivities u_1 and u_2, with time τ for different cases IH, IF, AH, and AF. It was shown from these figures that the anisotropy and functionally graded material have great effects on the nonlinear displacement sensitivities through the FGA composite microstructure.

Figures 5–7 show the variation of the nonlinear thermal stress sensitivities σ_{11}, σ_{12}, and σ_{22}, respectively, with time τ for different cases IH, IF, AH, and AF. It was noted from these figures that the anisotropy and functionally graded material have

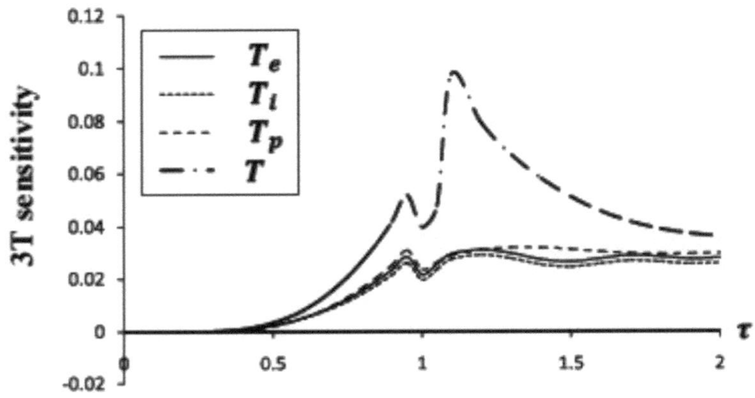

Figure 2.
Variation of the temperature sensitivity with time τ.

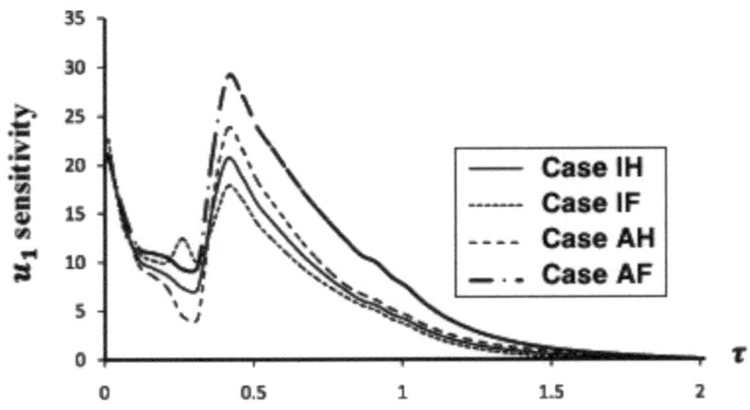

Figure 3.
Variation of the displacement u_1 sensitivity with time τ.

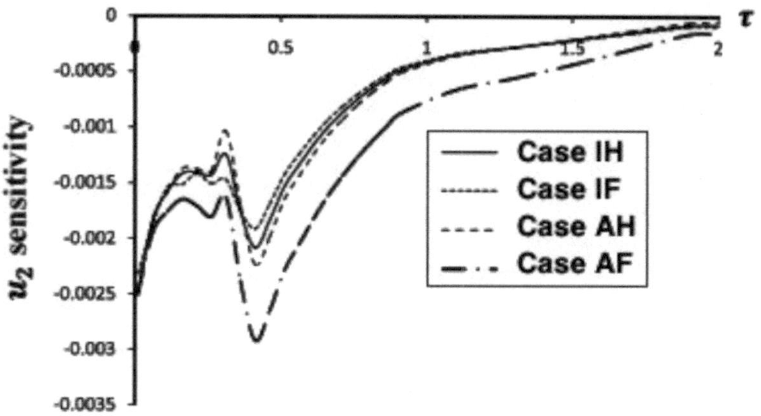

Figure 4.
Variation of the displacement u_2 sensitivity with time τ.

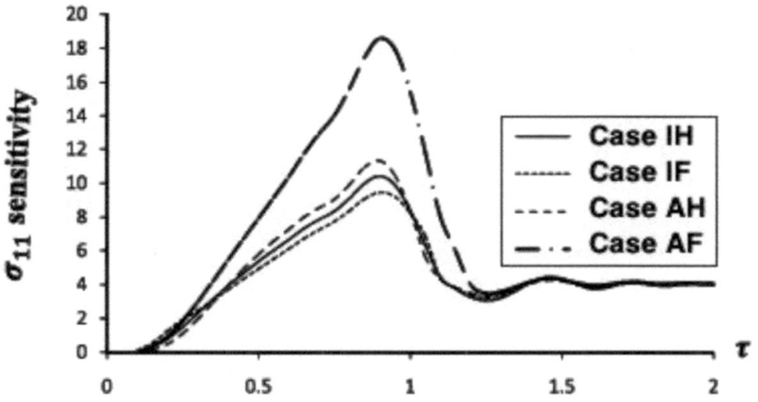

Figure 5.
Variation of the thermal stress σ_{11} sensitivity with time τ.

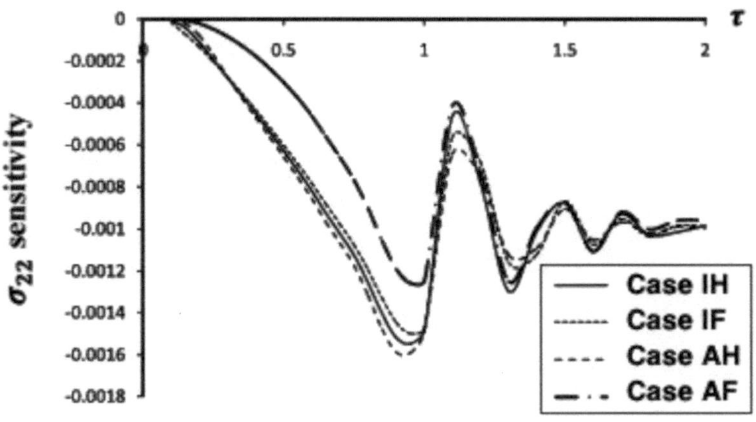

Figure 6.
Variation of the thermal stress σ_{12} sensitivity with time τ.

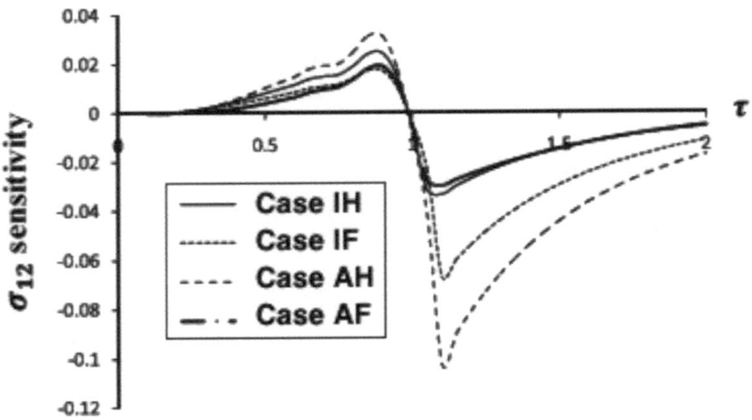

Figure 7.
Variation of the thermal stress σ_{22} sensitivity with time τ.

great influences on the nonlinear thermal stress sensitivities through the FGA composite microstructure.

For comparison purposes with those of other studies, we only considered one-dimensional numerical results of the considered three-temperature problem. In the considered special case, the nonlinear displacement u_1 and nonlinear thermal stress σ_{11} results are plotted in **Figures 8** and **9**, respectively. It can be noticed from these that the BEM results, which are based on replacing one-temperature heat conduction with three-temperature heat conduction, are in excellent agreement when compared to results obtained from the finite difference method of Pazera and Jędrysiak [97] and the finite element method (FEM) of Xiong and Tian [98]. We thus demonstrate the validity and accuracy of our proposed BEM technique.

Three numerical examples of BESO topological optimization of composite microstructures are performed to illustrate the optimization results of this study [99]. In order to obtain the functionally graded parameter effects during the optimization process of the considered composite microstructure, we consider the following values $m = 0, 0.5, 0.75$, and 1 in the one-temperature heat conduction model and the three-temperature radiative heat conduction model.

Example 1. Composite microstructures without holes or inclusions.

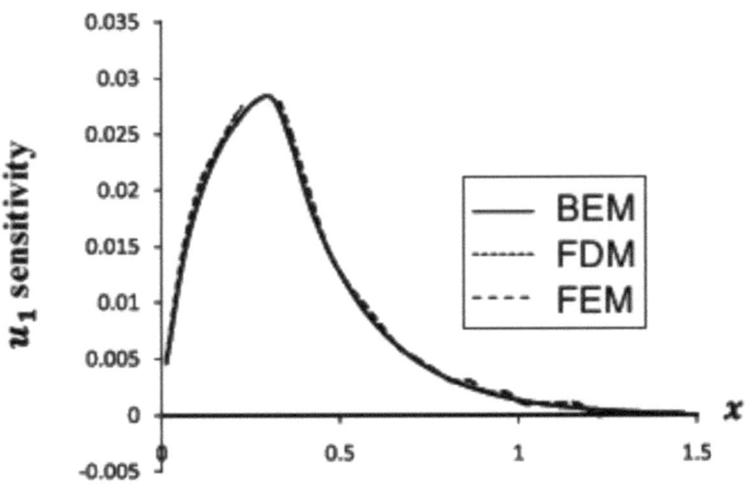

Figure 8.
Variation of the displacement u_1 sensitivity along x-axis.

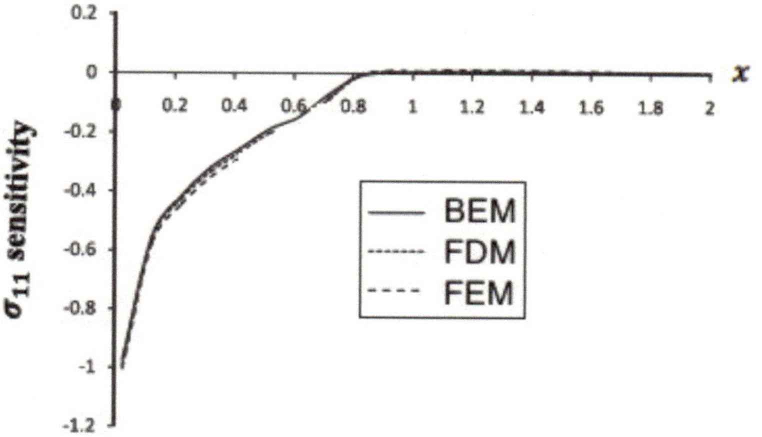

Figure 9.
Variation of the thermal stress σ_{11} sensitivity along x-axis.

The mean compliance has been minimized, to obtain the maximum stiffness for the composite microstructures made from two competitive materials and without holes or inclusions. Investigation of the effect of the functionally graded parameter on the optimal composite microstructure has been shown in **Table 1** for the $1T$ model and in **Table 2** for the $3T$ model. It is noticed from these tables that the heat conduction model and functionally graded parameter have a significant effect on the topology optimization process of the multi-material FGA composite microstructures.

Example 2. Composite microstructures with circular or square holes.

The mean compliance has been minimized to obtain the maximum stiffness for the composite microstructures made from two competitive materials and with circular or square holes. Investigation of the effect of the functionally graded

m	0	0.50	0.75	1.0
Design	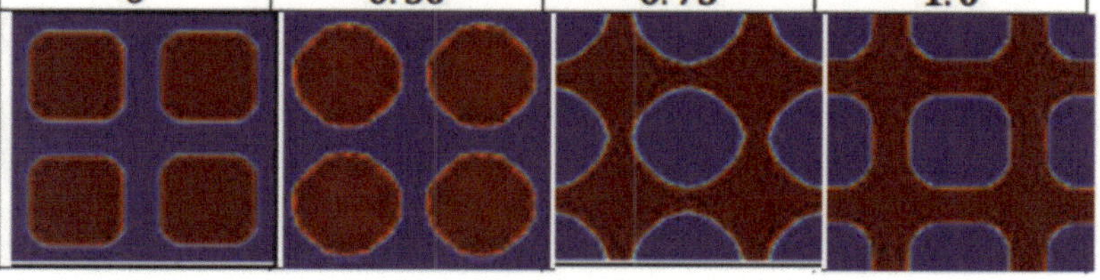			

Table 1.
Investigation of the influence of functionally graded parameter, m, *on the optimal composite microstructure for the* 1T *model.*

m	0	0.50	0.75	1.0
Design	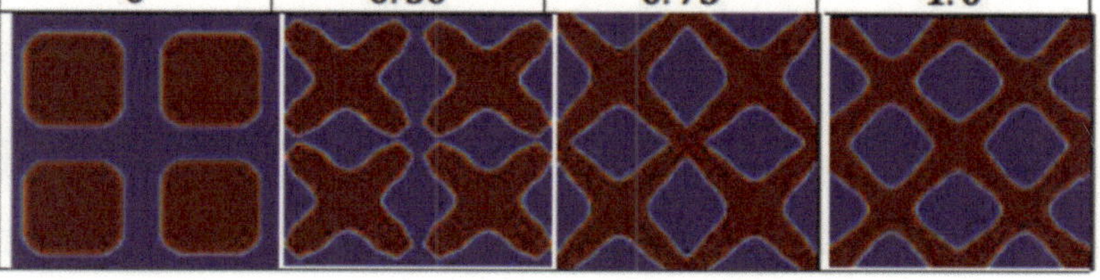			

Table 2.
Investigation of the influence of functionally graded parameter, m, *on the optimal composite microstructure for the* $3T$ *model.*

parameter on the optimal composite microstructure with circular holes has been shown in **Table 3** for the 1*T* model and in **Table 4** for the 3*T* model. Also, the investigation of the effect of the functionally graded parameter on the optimal composite microstructure with square holes has been shown in **Table 5** for the 1*T* model and in **Table 6** for the 3*T* model. It is noticed from these tables that the heat

m	0	0.50	0.75	1.0
Design	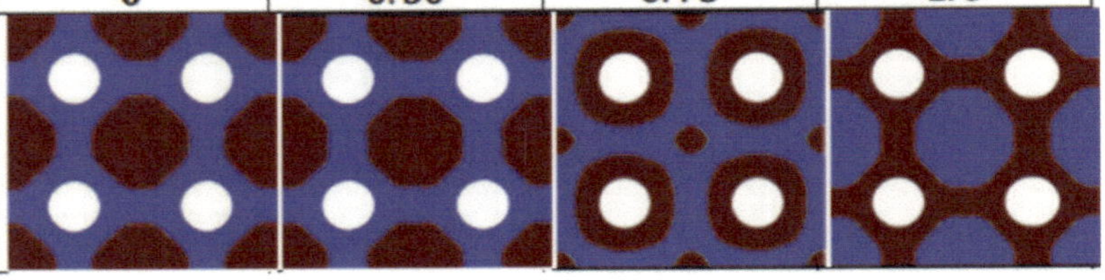			

Table 3.
Investigation of the influence of functionally graded parameter, m, *on the optimal composite microstructure with circular shape holes for the* 1T *model.*

m	0	0.50	0.75	1.0
Design	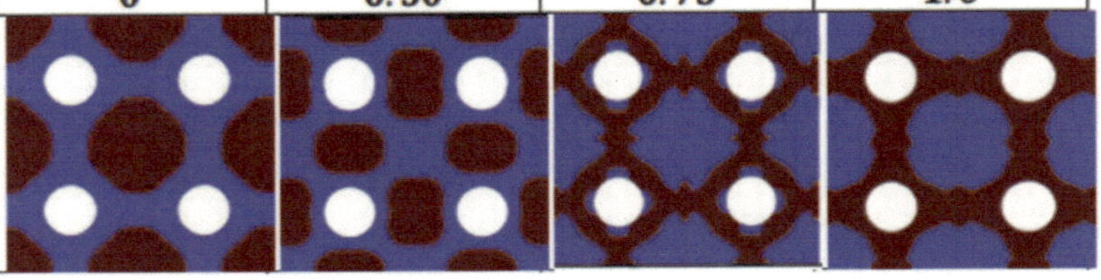			

Table 4.
Investigation of the influence of functionally graded parameter m on the optimal composite microstructure with circular shape holes for the 3T *model.*

m	0	0.50	0.75	1.0
Design	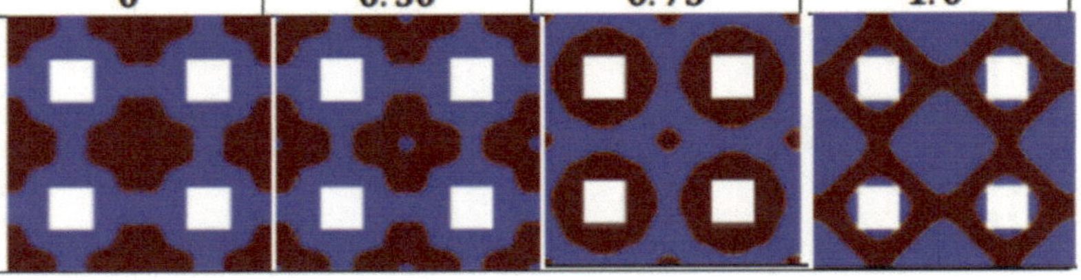			

Table 5.
Investigation of the influence of functionally graded parameter, m, *on the optimal composite microstructure with square shape holes for the* 1T *model.*

m	0	0.50	0.75	1.0
Design	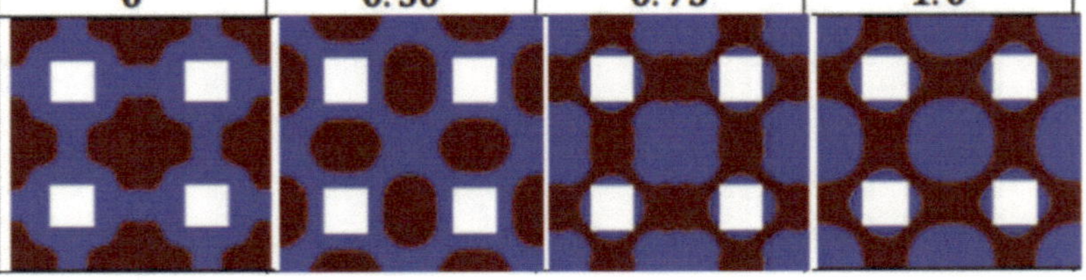			

Table 6.
Investigation of the influence of functionally graded parameter, m, *on the optimal composite microstructure with square shape holes for the* 3T *model.*

conduction model, functionally graded parameter, and holes shape have a significant effect on the topology optimization process of the multi-material FGA composite microstructures.

Example 3. Composite microstructures with circular or square inclusions.

The mean compliance has been minimized to obtain the maximum stiffness for the composite microstructures made from two competitive materials and with circular or square inclusions. Investigation of the effect of the functionally graded parameter on optimal composite microstructure with circular inclusions has been shown in **Table 7** for the $1T$ model and in **Table 8** for the $3T$ model. Also, the investigation of the effect of the functionally graded parameter on the optimal composite microstructure with square inclusions has been shown in **Table 9** for the $1T$ model and in **Table 10** for the $3T$ model. It is noticed from these tables that the heat conduction model, functionally graded parameter, and inclusions shape have a significant effect on the topology optimization process of the multi-material FGA composite microstructures.

The BESO topology optimization problem implemented in the numerical examples to find the distribution of the two materials in the design domain that minimize

m	0	0.50	0.75	1.0
Design	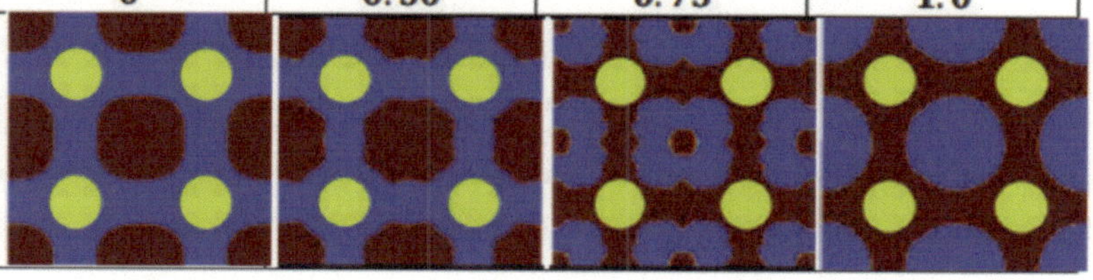			

Table 7.
Investigation of the influence of functionally graded parameter, m*, on the optimal composite microstructure with circular shape inclusions for the* 1T *model.*

m	0	0.50	0.75	1.0
Design	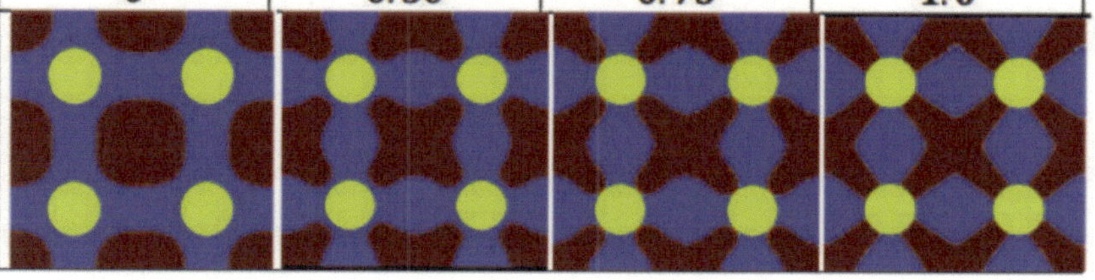			

Table 8.
Investigation of the influence of functionally graded parameter, m, on the optimal composite microstructure with circular shape inclusions for the 3T *model.*

m	0	0.50	0.75	1.0
Design	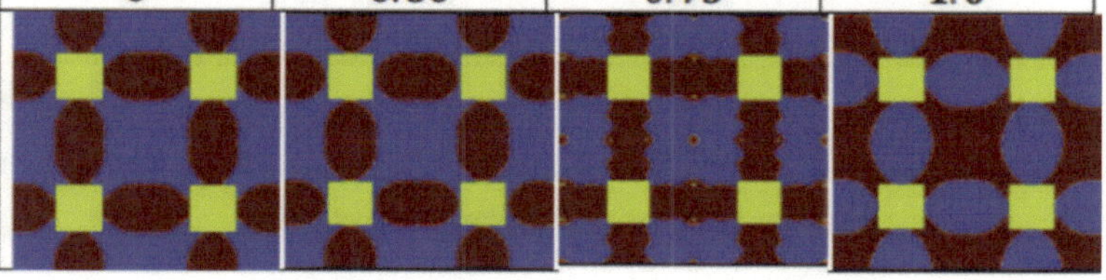			

Table 9.
Investigation of the influence of the functionally graded parameter, m*, on the optimal composite microstructure with square shape inclusions for the 1T model.*

m	0	0.50	0.75	1.0
Design	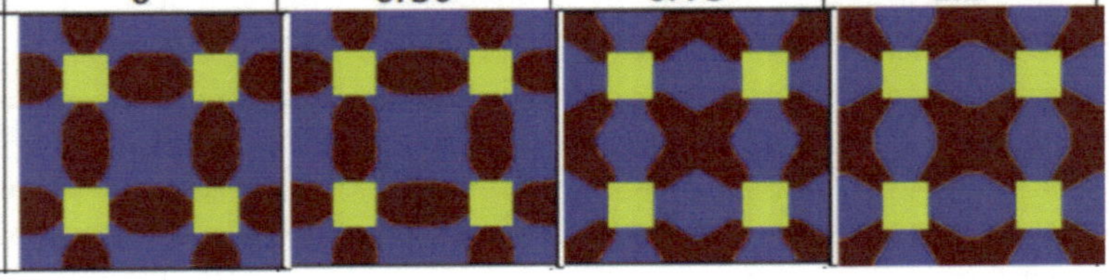			

Table 10.
Investigation of the influence of functionally graded parameter, m, *on the optimal composite microstructure with square shape inclusions for the* 3T *model.*

the compliance of the structure subject to a volume constraint in both phases can be stated as

Find X^M

That minimize $C^M = \frac{1}{2}\left(P^M\right)^T u^M = \frac{1}{2}\left(f^{M,ter} + f^{M,mec}\right)^T u^M$

Subject to $V_j^{M,*} - \Sigma_{i=1}^{N} V_i^M X_{ij}^M - \Sigma_{i=1}^{j-1} V_i^{M,*} = 0; j = 1, 2$

$$K^M u^M = P^M$$

$$X_i^M = x_{min} \mathrm{V}1; j = 1, 2$$

where X^M is the design variable; $V_j^{M,*}$ is the volume of the jth material phase, where i and j denote the element ith which is made of jth material; C^M is the mean compliance; P is the total load on the structure, which is the sum of mechanical and thermal loads; u^M is the displacement vector; $V^{M,*}$ is the volume of the solid material; N is the total number of elements; K^M is the global stiffness matrix; x_{min} is a small value (e.g., 0.0001), which guarantees that none of the elements will be removed completely from design domain; $f^{M,mec}$ is the mechanical load vector; and $f^{M,ter}$ is the thermal load vector. Also, the BESO parameters considered in these examples can be seen in **Table 11**. The validity of our implemented BESO topology optimization technique has been demonstrated in our recent reference [100].

Variable name	Variable description	Variable value
V_{f1}^M	Final volume fraction of the material 1 for both interpolations	0.10
V_{f2}^M	Final volume fraction of the material 2 for both interpolations	0.20
ER^M	Evolutionary ratio for interpolation 1	2%
ER^M	Evolutionary ratio for interpolation 2	3%
AR_{max}^M	Volume addition ratio for interpolation 1	3%
AR_{max}^M	Volume addition ratio for interpolation 2	2%
r_{min}^M	Filter ratio for interpolation 1	4 mm
r_{min}^M	Filter ratio for interpolation 2	3 mm
τ	Convergence tolerance for both interpolations	0.01%
N	Convergence parameter for both interpolations	5

Table 11.
Multi-material BESO parameters for minimization of a composite microstructure.

Model	Type	Method	σ_{11} sensitivity	σ_{12} sensitivity	σ_{22} sensitivity
		BEM (present)	0.4084297	0.0509346	0.5332620
1T	SS	FEM [101]	0.4084297	0.0509346	0.5332620
		FVM [102]	0.4084297	0.0509346	0.5332620
		BEM (present)	0.3591487	0.0408259	0.3758618
1T	CC	FEM [101]	0.3591487	0.0408259	0.3758618
		FVM [102]	0.3591487	0.0408259	0.3758618
		BEM (present)	0.2518379	0.0307736	0.2613532
1T	CS	FEM [101]	0.2518378	0.0307735	0.2613531
		FVM [102]	0.2518379	0.0307736	0.2613532
		BEM (present)	0.3147697	0.0304365	0.4767924
3T	SS	FEM [101]	0.3147696	0.0304364	0.4767923
		FVM [102]	0.3147697	0.0304365	0.4767924
		BEM (present)	0.2432756	0.0204748	0.3052857
3T	CC	FEM [101]	0.2432755	0.0204747	0.3052856
		FVM [102]	0.2432756	0.0204748	0.3052857
		BEM (present)	0.1258948	0.0107825	0.2079735
3T	CS	FEM [101]	0.1258947	0.0107824	0.2079734
		FVM [102]	0.1258948	0.0107825	0.2079737

Table 12.
Models of 1T and 3T nonlinear thermal stresses' sensitivities for different types of boundary conditions and different methods.

Example 4. Laminated composite microstructure with three different sets of boundary conditions are considered in this example to validate the BEM formulation of the current study. These boundary conditions are called: simply—simply supported (SS), clamped—clamped (CC), and clamped—simply supported (CS). One-temperature (1T) and three-temperature (3T) models of nonlinear thermal stresses sensitivities results have been compared with the finite element method (FEM) results of Rajanna et al. [101] as well as with the finite volume method (FVM) results of Fallah and Delzendeh [102], which are tabulated in **Table 12** for different types of boundary conditions and different methods. It can be observed that the BEM results for all the three types of boundary conditions are in excellent agreement with FEM results of [101] and the FVM results of [102].

6. Conclusion

The main aim of this chapter is to describe a new boundary element formulation for the modeling and optimization of the three-temperature nonlinear generalized magneto-thermoelastic functionally graded anisotropic (FGA) composite microstructures. The governing equations of the considered model are very difficult to solve analytically because of the nonlinearity and anisotropy. To overcome this, we propose a new boundary element formulation for solving such equations, where we used the three-temperature nonlinear radiative heat conduction equations combined with electron, ion, and phonon temperatures. Numerical results show the three-temperature distributions through composite microstructure. The effects of anisotropy and functionally graded material on the three-temperature nonlinear displacement sensitivities and nonlinear thermal stress sensitivities through the composite microstructure are very significant and pronounced. Because there are no available results in the literature to confirm the validity and accuracy of our proposed technique except for one-temperature heat conduction, we replace the three-temperature radiative heat conduction with one-temperature heat conduction as a special case from our current general study. In the considered special case, the BEM results have been compared graphically with the FDM results and FEM results, and it can be noticed that the BEM results are in excellent agreement with the FDM and FEM results. These results thus demonstrate the validity and accuracy of our proposed technique.

Numerical examples are solved using the method of moving asymptotes (MMA) algorithm based on the bi-evolutionary structural optimization method (BESO), where we used the topological optimization to manufacture three-temperature magneto-thermoelastic composite microstructures to obtain the required specific engineering properties. A new class of FGA composite microstructures consisting of two competitive materials has been studied, taking into account existence of holes or inclusions. The effects of the heat conduction model, functionally graded parameter, and holes shape and inclusions shape on the optimal composite microstructure are investigated through the considered examples with great practical interest.

The ability to understand and manipulate composite microstructures has been fundamental to our technical development over time. Today, scientists and engineers recognize the importance of composite microstructures use for economic and environmental reasons. Based on the BEM implementation and its results, this study concluded that the boundary element technique is the most suitable technique for the manufacturing of FGA composite microstructures in the future works. This technique aimed to describe the behavior of FGA composite microstructures and achieves improvement in the composition optimization and mechanical properties of the resulting FGA composite microstructures.

Due to three-temperature and numerous low-temperature and high-temperature applications in laminated composites microstructures, as a future work and based on the findings obtained in the present study, we would suggest further research to develop numerical techniques for solving the three-temperature nonlinear thermoelastic wave propagation problems and for manufacturing of advanced laminated composites. The numerical results of our considered study can provide data references for mechanical engineers, computer engineers, geotechnical engineers, geothermal engineers, technologists, new materials designers, physicists, material science researchers, and those who are interested in novel technologies in the area of three-temperature magneto-thermoelastic FGA composite microstructures. Application of three-temperature theories in advanced manufacturing technologies, with the development of soft machines and robotics in biomedical engineering and advanced manufacturing, and nonlinear generalized magneto-thermoelastic problems will be encountered more often where three-temperature radiative heat conduction will turn out to be the best choice for thermomechanical analysis in the design and analysis of advanced composite microstructures.

Author details

Mohamed Abdelsabour Fahmy[1,2]

1 Jamoum University College, Umm Al-Qura University, Jamoum, Makkah, Saudi Arabia

2 Faculty of Computers and Informatics, Suez Canal University, New Campus, Ismailia, Egypt

*Address all correspondence to: mohamed_fahmy@ci.suez.edu.eg

References

[1] Pindera MJ, Arnold SM, Aboudi J, Hui D. Use of composites in functionally graded materials. Composites Engineering. 1994;**4**:1-145

[2] Pindera MJ, Aboudi J, Arnold SM, Jones WF. Use of composites in multi-phased and functionally graded materials. Composites Engineering. 1995;**5**:743-974

[3] Yin HM, Paulino GH, Buttlar WG, Sun LZ. Effective thermal conductivity of two-phase functionally graded particulate composites. Journal of Applied Physics. 2005;**98**:063704

[4] Miyamoto Y, Kaysser WA, Rabin BH, Kawasaki A, Ford RG. Functionally Graded Materials: Design, Processing and Applications. New York: Springer US; 1999

[5] Noda N. Thermal stresses in functionally graded material. Journal of Thermal Stresses. 1999;**22**:477-512

[6] Kieback B, Neubrand A, Riedel H. Processing techniques for functionally graded materials. Materials Science and Engineering. 2003;**362**:81-106

[7] Kawasaki A, Watanabe R. Microstructural designing and fabrication of disk shaped functionally gradient materials by powder metallurgy. Journal of the Japan Society of Power and Powder Metallurgy. 1990; **37**:253-258

[8] Kiebact B, Neubrand A. Processing techniques for functionally graded materials. Materials Science and Engineering A. 2003;**362**:81-85

[9] Fahmy MA. A time-stepping DRBEM for 3D anisotropic functionally graded piezoelectric structures under the influence of gravitational waves. In: Rodrigues H, Elnashai A, Calvi G. editors. Facing the Challenges in Structural Engineering. Sustainable Civil Infrastructures. 15-19 July 2017; Sharm El Sheikh, Egypt (GeoMEast 2017). Cham: Springer; 2018. pp. 350-365. DOI: 10.1007/978-3-319-61914-9_27

[10] Fahmy MA. 3D DRBEM modeling for rotating initially stressed anisotropic functionally graded piezoelectric plates. In: Proceedings of the 7th European Congress on Computational Methods in Applied Sciences and Engineering (ECCOMAS 2016); 5-10 June 2016; Crete Island, Greece. 2016. pp. 7640-7658

[11] Fahmy MA. Boundary element solution of 2D coupled problem in anisotropic piezoelectric FGM plates. In: Proceedings of the 6th International Conference on Computational Methods for Coupled Problems in Science and Engineering (Coupled Problems 2015); 18-20 May 2015; Venice, Italy. 2015. pp. 382-391

[12] Fahmy MA. The DRBEM solution of the generalized magneto-thermo-viscoelastic problems in 3D anisotropic functionally graded solids. In: Proceedings of the 5th International Conference on Coupled Problems in Science and Engineering (Coupled Problems 2013); 17-19 June 2013; Ibiza, Spain. 2013. pp. 862-872

[13] Fahmy MA. Computerized Boundary Element Solutions for Thermoelastic Problems: Applications to Functionally Graded Anisotropic Structures. Saarbrücken: LAP Lambert Academic Publishing; 2017

[14] Fahmy MA. Boundary Element Computation of Shape Sensitivity and Optimization: Applications to Functionally Graded Anisotropic Structures. Saarbrücken: LAP Lambert Academic Publishing; 2017

[15] Fahmy MA. A computerized DRBEM model for generalized magneto-thermo-visco-elastic stress waves in functionally graded anisotropic thin film/substrate structures. Latin American Journal of Solids and Structures. 2014;**11**:386-409

[16] Fahmy MA, Salem AM, Metwally MS, Rashid MM. Computer implementation of the DRBEM for studying the classical coupled thermoelastic responses of functionally graded anisotropic plates. Physical Science International Journal. 2014;**4**: 674-685

[17] Fahmy MA, Salem AM, Metwally MS, Rashid MM. Computer implementation of the DRBEM for studying the generalized thermo elastic responses of functionally graded anisotropic rotating plates with two relaxation times. British Journal of Mathematics & Computer Science. 2014;**4**:1010-1026

[18] Fahmy MA. DRBEM sensitivity analysis and shape optimization of rotating magneto-thermo-viscoelastic FGA structures using golden-section search algorithm based on uniform bicubic B-splines. Journal of Advances in Mathematics and Computer Science. 2017;**25**:1-20

[19] Fahmy MA. A predictor-corrector time-stepping DRBEM for shape design sensitivity and optimization of multilayer FGA structures. Transylvanian Review. 2017;**XXV**: 5369-5382

[20] Fahmy MA, Al-Harbi SM, Al-Harbi BH, Sibih AM. A computerized boundary element algorithm for modeling and optimization of complex magneto-thermoelastic problems in MFGA structures. Journal of Engineering Research and Reports. 2019;**3**:1-13

[21] Fahmy MA. A new LRBFCM-GBEM modeling algorithm for general solution of time fractional order dual phase lag bioheat transfer problems in functionally graded tissues. Numerical Heat Transfer, Part A: Applications. 2019;**75**:616-626

[22] Hyun S, Torquato S. Designing composite microstructures with targeted properties. Journal of Materials Research. 2001;**16**:280-285

[23] Rodriguez R, Kelestemur MH. Processing and microstructural characterization of functionally gradient Al A356/SiCp composite. Journal of Materials Science. 2002;**37**:1813-1821

[24] Duhamel J. Some memoire sur les phenomenes thermo-mechanique. Journal de l'École polytechnique. 1837; **15**:1-57

[25] Neumann F. Vorlesungen Uber die theorie der elasticitat. Meyer: Brestau; 1885

[26] Biot M. Thermoelasticity and irreversible thermo-dynamics. Journal of Applied Physics. 1956;**27**:249-253

[27] Lord HW, Shulman Y. A generalized dynamical theory of thermoelasticity. Journal of the Mechanics and Physics of Solids. 1967;**15**:299-309

[28] Green AE, Lindsay KA. Thermoelasticity. Journal of Elasticity. 1972;**2**:1-7

[29] Green AE, Naghdi PM. On undamped heat waves in an elastic solid. Journal of Thermal Stresses. 1992;**15**: 253-264

[30] Green AE, Naghdi PM. Thermoelasticity without energy dissipation. Journal of Elasticity. 1993; **31**:189-208

[31] Tzou DY. A unified field approach for heat conduction from macro to micro scales. ASME Journal of Heat Transfer. 1995;**117**:8-16

[32] Chandrasekharaiah DS. Hyperbolic thermoelasticity: A review of recent literature. Applied Mechanics Reviews. 1998;**51**:705-729

[33] Roychoudhuri SK. On a thermoelastic three-phase-lag model. Journal of Thermal Stresses. 2007;**30**: 231-238

[34] Fahmy MA. A time-stepping DRBEM for magneto-thermo-viscoelastic interactions in a rotating nonhomogeneous anisotropic solid. International Journal of Applied Mechanics. 2011;**3**:1-24

[35] Fahmy MA. A time-stepping DRBEM for the transient magneto-thermo-visco-elastic stresses in a rotating non-homogeneous anisotropic solid. Engineering Analysis with Boundary Elements. 2012;**36**:335-345

[36] Fahmy MA. Numerical modeling of transient magneto-thermo-viscoelastic waves in a rotating nonhomogeneous anisotropic solid under initial stress. International Journal of Modeling. Simulation and Scientific Computing. 2012;**3**:1250002

[37] Fahmy MA. Transient magneto-thermo-viscoelastic stresses in a rotating nonhomogeneous anisotropic solid with and without a moving heat source. Journal of Engineering Physics and Thermophysics. 2012;**85**:950-958

[38] Fahmy MA. Transient magneto-thermo-elastic stresses in an anisotropic viscoelastic solid with and without moving heat source. Numerical Heat Transfer Part A: Applications. 2012;**61**: 547-564

[39] Fahmy MA. Transient magneto-thermoviscoelastic plane waves in a non-homogeneous anisotropic thick strip subjected to a moving heat source. Applied Mathematical Modelling. 2012; **36**:4565-4578

[40] Fahmy MA. The effect of rotation and inhomogeneity on the transient magneto-thermoviscoelastic stresses in an anisotropic solid. ASME Journal of Applied Mechanics. 2012;**79**: 1015

[41] Sharma N, Mahapatra TR, Panda SK. Thermoacoustic behavior of laminated composite curved panels using higher-order finite-boundary element model. International Journal of Applied Mechanics. 2018;**10**:1850017

[42] Othman MIA, Khan A, Jahangir R, Jahangir A. Analysis on plane waves through magneto-thermoelastic microstretch rotating medium with temperature dependent elastic properties. Applied Mathematical Modelling. 2019;**65**:535-548

[43] Ezzat MA, El-Karamany AS, El-Bary AA. On dual-phase-lag thermoelasticity theory with memory-dependent derivative. Mechanics of Advanced Materials and Structures. 2017;**24**:908-916

[44] Ezzat MA, El-Karamany AS, El-Bary AA. Generalized thermoelasticity with memory-dependent derivatives involving two temperatures. Mechanics of Advanced Materials and Structures. 2016;**23**:545-553

[45] Fahmy MA. A computerized boundary element model for simulation and optimization of fractional-order three temperatures nonlinear generalized piezothermoelastic problems based on genetic algorithm. In: AIP Conference Proceedings 2138 of Innovation and Analytics Conference and Exihibiton (IACE 2019); 25-28 March 2019; Sintok, Malaysia. 2019. p. 030015

[46] Fahmy MA. A new computerized boundary element model for three-temperature nonlinear generalized thermoelastic stresses in anisotropic circular cylindrical plate

structures. In: Awrejcewicz J, Grzelczyk D, editors. Dynamical Systems Theory. London, UK: IntechOpen; 2019. pp. 1-17

[47] Fahmy MA. Boundary element model for nonlinear fractional-order heat transfer in magneto-thermoelastic FGA structures involving three temperatures. In: Ebrahimi F, editor. Mechanics of Functionally Graded Materials and Structures. London, UK: IntechOpen; 2019. pp. 1-22

[48] Fahmy MA. Boundary element mathematical modelling and boundary element numerical techniques for optimization of micropolar thermoviscoelastic problems in solid deformable bodies. In: Sivasankaran S, Nayak PK, Günay E, editors. Mechanics of Solid Deformable Bodies. London, UK: IntechOpen; 2020. pp. 1-21

[49] Fahmy MA. Boundary element modeling and optimization based on fractional-order derivative for nonlinear generalized photo-thermoelastic stress wave propagation in three-temperature anisotropic semiconductor structures. In: Sadollah A, Sinha TS, editors. Recent Trends in Computational Intelligence. London, UK: IntechOpen; 2020. pp. 1-16

[50] El-Naggar AM, Abd-Alla AM, Fahmy MA, Ahmed SM. Thermal stresses in a rotating non-homogeneous orthotropic hollow cylinder. Heat and Mass Transfer. 2002;**39**:41-46

[51] El-Naggar AM, Abd-Alla AM, Fahmy MA. The propagation of thermal stresses in an infinite elastic slab. Applied Mathematics and Computation. 2003;**12**:220-226

[52] Abd-Alla AM, El-Naggar AM, Fahmy MA. Magneto-thermoelastic problem in non-homogeneous isotropic cylinder. Heat and Mass Transfer. 2003; **39**:625-629

[53] Hu Q, Zhao L. Domain decomposition preconditioners for the system generated by discontinuous Galerkin discretization of 2D-3T heat conduction equations. Communications in Computational Physics. 2017;**22**: 1069-1100

[54] Cho JR, Ha DY. Averaging and finite element discretization aproaches in the numerical analysis of functionally graded materials. Materials Science and Engineering A. 2001;**302**:187-196

[55] Sharma N, Mahapatra TR, Panda SK. Thermoacoustic behavior of laminated composite curved panels using higher-order finite-boundary element model. International Journal of Applied Mechanics. 2018;**10**:1850017

[56] Eskandari AH, Baghani M, Sohrabpour S. A time-dependent finite element formulation for thick shape memory polymer beams considering shear effects. International Journal of Applied Mechanics. 2019;**10**: 1850043

[57] Soliman AH, Fahmy MA. Range of applying the boundary condition at fluid/porous interface and evaluation of beavers and Joseph's slip coefficient using finite element method. Computation. 2020;**8**:14

[58] Fahmy MA. A new boundary element strategy for modeling and simulation of three temperatures nonlinear generalized micropolar-magneto-thermoelastic wave propagation problems in FGA structures. Engineering Analysis with Boundary Elements. 2019;**108**:192-200

[59] Fahmy MA. A three-dimensional generalized magneto-thermo-viscoelastic problem of a rotating functionally graded anisotropic solids with and without energy dissipation. Numerical Heat Transfer, Part A: Applications. 2013;**63**:713-733

[60] Fahmy MA. A 2-D DRBEM for generalized magneto-thermo-viscoelastic transient response of rotating functionally graded anisotropic thick strip. International Journal of Engineering and Technology Innovation. 2013;**3**:70-85

[61] Fahmy MA, Salem AM, Metwally MS, Rashid MM. Computer implementation of the DRBEM for studying the generalized thermoelastic responses of functionally graded anisotropic rotating plates with one relaxation time. International Journal of Applied Science and Technology. 2013; **3**:130-140

[62] Fahmy MA, Salem AM, Metwally MS, Rashid MM. Computer implementation of the DRBEM for studying the classical uncoupled theory of thermoelasticity of functionally graded anisotropic rotating plates. International Journal of Engineering Research and Applications. 2013;**3**: 1146-1154

[63] Fahmy MA. A Computerized Boundary Element Models for Coupled, Uncoupled and Generalized Thermoelasticity Theories of Functionally Graded Anisotropic Rotating Plates. UK: Book Publisher International; 2019

[64] Fahmy MA. A new computerized boundary element algorithm for cancer modeling of cardiac anisotropy on the ECG simulation. Asian Journal of Research in Computer Science. 2018;**2**: 1-10

[65] Brebbia CA, Telles JCF, Wrobel L. Boundary Element Techniques in Engineering. New York: Springer-Verlag; 1984

[66] Wrobel LC, Brebbia CA. The dual reciprocity boundary element formulation for nonlinear diffusion problems. Computer Methods in Applied Mechanics and Engineering. 1987;**65**:147-164

[67] Partridge PW, Brebbia CA. Computer implementation of the BEM dual reciprocity method for the solution of general field equations. Communications in Applied Numerical Methods. 1990;**6**:83-92

[68] Partridge PW, Brebbia CA, Wrobel LC. The Dual Reciprocity Boundary Element Method. Southampton: Computational Mechanics Publications; 1992

[69] Fahmy MA. Boundary element algorithm for nonlinear modeling and simulation of three temperature anisotropic generalized micropolar piezothermoelasticity with memory-dependent derivative. International Journal of Applied Mechanics. 2020;**12**: 2050027

[70] Abd-Alla AM, Fahmy MA, El-Shahat TM. Magneto-thermo-elastic problem of a rotating non-homogeneous anisotropic solid cylinder. Archive of Applied Mechanics. 2008;**78**:135-148

[71] Fahmy MA. A New BEM for Modeling and Simulation of Laser Generated Ultrasound Waves in 3T Fractional Nonlinear Generalized Micropolar Poro-Thermoelastic FGA Structures. In: Valdman J, Marcinkowski L, editors. Modeling and Simulation in Engineering. London, UK: IntechOpen; 2020

[72] Fahmy MA. Implicit-explicit time integration DRBEM for generalized magneto-thermoelasticity problems of rotating anisotropic viscoelastic functionally graded solids. Engineering Analysis with Boundary Elements. 2013; **37**:107-115

[73] Fahmy MA. Generalized magneto-thermo-viscoelastic problems of rotating functionally graded anisotropic plates by the dual reciprocity boundary element method. Journal of Thermal Stresses. 2013;**36**:1-20

[74] Fahmy MA. A 2D time domain DRBEM computer model for magneto-thermoelastic coupled wave propagation problems. International Journal of Engineering and Technology Innovation. 2014;**4**:138-151

[75] Fahmy MA, Al-Harbi SM, Al-Harbi BH. Implicit time-stepping DRBEM for design sensitivity analysis of magneto-thermo-elastic FGA structure under initial stress. American Journal of Mathematical and Computational Sciences. 2017;**2**:55-62

[76] Fahmy MA. The effect of anisotropy on the structure optimization using golden-section search algorithm based on BEM. Journal of Advances in Mathematics and Computer Science. 2017;**25**:1-18

[77] Fahmy MA. Shape design sensitivity and optimization of anisotropic functionally graded smart structures using bicubic B-splines DRBEM. Engineering Analysis with Boundary Elements. 2018;**87**:27-35

[78] Fahmy MA. Shape design sensitivity and optimization for two-temperature generalized magneto-thermoelastic problems using time-domain DRBEM. Journal of Thermal Stresses. 2018;**41**:119-138

[79] Fahmy MA. Boundary element algorithm for modeling and simulation of dual-phase lag bioheat transfer and biomechanics of anisotropic soft tissues. International Journal of Applied Mechanics. 2018;**10**:1850108

[80] Fahmy MA. Modeling and optimization of anisotropic viscoelastic porous structures using CQBEM and moving asymptotes algorithm. Arabian Journal for Science and Engineering. 2019;**44**:1671-1684

[81] Fahmy MA. Boundary element modeling and simulation of biothermomechanical behavior in anisotropic laser-induced tissue hyperthermia. Engineering Analysis with Boundary Elements. 2019;**101**:156-164

[82] Fahmy MA. Design optimization for a simulation of rotating anisotropic viscoelastic porous structures using time-domain OQBEM. Mathematics and Computers in Simulation. 2019;**66**:193-205

[83] Fahmy MA. A new convolution variational boundary element technique for design sensitivity analysis and topology optimization of anisotropic thermo-poroelastic structures. Arab Journal of Basic and Applied Sciences. 2020;**27**:1-12

[84] Fahmy MA. Thermoelastic stresses in a rotating non-homogeneous anisotropic body. Numerical Heat Transfer, Part A: Applications. 2008;**53**:1001-1011

[85] Fahmy MA, El-Shahat TM. The effect of initial stress and inhomogeneity on the thermoelastic stresses in a rotating anisotropic solid. Archive of Applied Mechanics. 2008;**78**:431-442

[86] Farhat C, Park KC, Dubois-Pelerin Y. An unconditionally stable staggered algorithm for transient finite element analysis of coupled thermoelastic problems. Computer Methods in Applied Mechanics and Engineering. 1991;**85**:349-365

[87] Svanberg K. The method of moving asymptotes a new method for structural optimization. International Journal of Numerical Methods in Engineering. 1987;**24**:359-373

[88] Huang X, Xie Y. Convergent and mesh-independent solutions for the bi-directional evolutionary structural optimization method. Finite Elements in

Analysis and Design. 2007;**43**(14): 1039-1049

[89] Huang X, Xie Y. Evolutionary Topology Optimization of Continuum Structures. USA: John Wiley & Sons Ltd.; 2010

[90] Huang X, Xie YM. Bi-directional evolutionary topology optimization of continuum structures with one or multiple materials. Computational Mechanics. 2008;**43**(3):393

[91] Huang X, Zhou S, Xie Y, Li Q. Topology optimization of microstructures of cellular materials and composites for macrostructures. Computational Materials Science. 2013; **67**:397-407

[92] Sigmund O. Design of multiphysics actuators using topology optimization - Part I: One material structures. Computer Methods in Applied Mechanics and Engineering. 2001; **190**(49):6577-6604

[93] Sigmund O, Torquato S. Composites with extremal thermal expansion coefficients. Applied Physics Letters. 1996;**69**(21):3203-3205

[94] Sigmund O, Torquato S. Design of materials with extreme thermal expansion using a three-phase topology optimization method. Journal of the Mechanics and Physics of Solids. 1997; **45**(6):1037-1067

[95] Wang Y, Luo Z, Zhang N, Wu T. Topological design for mechanical metamaterials using a multiphase level set method. Structural and Multidisciplinary Optimization. 2016b; **54**:937-954

[96] Xu B, Huang X, Zhou S, Xie Y. Concurrent topological design of composite thermoelastic macrostructure and microstructure with multi-phase material for maximum stiffness. Composite Structures. 2016;**150**:84-102

[97] Pazera E, Jędrysiak J. Effect of microstructure in thermoelasticity problems of functionally graded laminates. Composite Structures. 2018; **202**:296-303

[98] Xiong QL, Tian XG. Generalized magneto-thermo-microstretch response during thermal shock. Latin American Journal of Solids and Structures. 2015; **12**:2562-2580

[99] Krysko AV, Awrejcewicz J, Pavlov SP, Bodyagina KS, Krysko VA. Topological optimization of thermoelastic composites with maximized stiffness and heat transfer. Composites Part B Engineering. 2019; **158**:319-327

[100] Fahmy MA. A new BEM for modeling and optimization of 3T fractional nonlinear generalized magneto-thermoelastic multi-material ISMFGA structures subjected to moving heat source. In: Koprowski R, editor. Fractal Analysis. London, UK: IntechOpen; 2020

[101] Rajanna T, Banerjee S, Desai YM, Prabhakara DL. Effect of boundary conditions and non-uniform edge loads on buckling characteristics of laminated composite panels with and without cutout. International Journal for Computational Methods in Engineering Science and Mechanics. 2017;**18**:64-76

[102] Fallah N, Delzendeh M. Free vibration analysis of laminated composite plates using meshless finite volume method. Engineering Analysis with Boundary Elements. 2018;**88**: 132-144